W9-BQV-692

LIFE
ON THE DRY LINE

LIFE
ON THE DRY LINE

WORKING THE LAND, 1902–1944

HARRY MORGAN MASON

FOREWORD
BY WES JACKSON

Fulcrum Publishing
Golden, Colorado

Library of Congress Cataloging-in-Publication Data

Mason, Harry Morgan
        Life on the dry line : working the land, 1902–1944 /
    Harry Morgan Mason ; foreword by Wes Jackson.
                p.   cm.
        ISBN 1–55591–122–6
        1. Farm life—Kansas—WaKeeney Region—History—
    20th century. 2. Agriculture—Kansas—WaKeeney Region—
    History—20th century. 3. Farm mechanization—Kansas—
    WaKeeney Region—History—20th century. 4. Mason, Harry
    Morgan. 5. WaKeeney Region (Kan.)—History. I. Title.
    S521.5.K2M27 1992
    630' .9781'16509041—dc20   92-53038
                                                            CIP

Printed in the United States of America

0   9   8   7   6   5   4   3   2   1

Fulcrum Publishing
350 Indiana Street, Suite 350
Golden, Colorado 80401-5093

To my sister, Mary Musgrave

CONTENTS

FOREWORD

by Wes Jackson

In the old, familiar story of westward expansion failure is a parenthetical event. If it is not Satan in Paradise Lost, then who wants to listen?
—Mary Beth LaDow

Harry Mason is an articulate mechanic. More than eighty years after he was born in 1908, he now describes what amounts to some of the important "nuts and bolts" of the industrial revolution which came to this near-last frontier in the lower forty-eight. He was born and raised smack-dab on the 100th meridian, halfway, almost to the mile, between Denver and Kansas City on the dry plains of Kansas. Mid-grass prairie gives way to short grass here or, coming from Denver, the other way around.

Plenty has been written about the railroads' penetration of the frontier, but the arrival of the car, truck and tractor to the farms and small towns of the frontier is a more fine-grained story. More sweepingly, it is the story of technological evolution where nature is subdued or ignored. In staggering detail, we learn of the most minor changes in those early-day engines—their fuel and ignition systems, their chasses and suspensions—destined to shape the current character of the continent.

This book has the potential to unite many audiences, from antique club and flywheel association members to professional historians. There is something for all of us to savor. As we view the various snapshots, we see the character of the people on that frontier and how motorized technology built at distant factories was ultimately at the mercy of those frontier mechanics. Between the lines we can see the ordinary weaknesses and strengths of the city manufacturers, government and "stay-put" settlers on the frontier.

The book goes beyond the details of the everyday tragedies which surely visited any small- to medium-sized town in America during that period. Even though Harry is a mechanic at heart, his first experiences were on his family farm which featured various animals, including draft animals. Here were the farms that became targets or, in more polite language, "one of the principal sectors of American life the industrial revolution set out to serve." This service included nearly eliminating that sun-powered creature, the draft animal. There was no simple industrial conspiracy. In fact, a more accurate hypothesis might be that the machine *was* Satan in a prairie paradise.

Harry took care of all kinds of farm livestock, including the draft animals. He has the courage to say, however, that he didn't like livestock very much. He empathized with those farmers more constitutionally predisposed to getting along with a tractor than a potentially runaway team of easily spooked horses. Those who loved machines benefited from a fossil fuel subsidy even though replacement for such power did not come in the form of a solar-produced foal but a four-wheeled mini-monster covered with green and yellow paint. A farmer who could get along with horses was up against more than a machine. He was up against a *subsidy* of cheap ancient energy, in seemingly inexhaustible amounts, and the glitziness of a paint job. But it is not that simple, either. It was also a matter of fashion that sent the horses and mules to the likes of the Hill Packing Company at Topeka to become pet food.

Enough speculation. At one point, Harry's father bought a garage in town. Here we see the young Harry Mason repairing the broken promises of the industrial revolution. He didn't do these repairs grudgingly. In fact, he seems to have repaired broken machines with enthusiasm. One gets the feeling those were the happiest days of his long life. It is touching to read what he felt, driving out in the country to fix a neighbor's tractor in the field, to read how gratifying it was for both the farmer and Harry to get the thing going again. It didn't matter that the hours were long, the winters cold and the summers hot and dry. Traditional virtues—patience, hard work, attention to detail and a good deal of love—went into helping the owners of those first machines of the era of industrialized agriculture.

Don't be misled. This is not a book about "Zen and the Art of Tractor and Automobile Maintenance" or the "Existential Pleasures of Engineering," though it clearly carries elements of both. It is a book written by a man who left the farm and garage and eventually earned a Ph.D. in experimental psychology. In spite of all his years in universities, university life was never as much fun as garage life.

The historians and others who want to know how it "felt on the farm" and around the garage in those days, the folk who want "less mechanical detail and more feeling" will have to develop the literacy to read between the lines.

There is no mention of Leo Marx's book *The Machine in the Garden*, a book about technology and the pastoral ideal in America. Nor does Harry mention *The Frontier in American History* essay published in 1893 by Harvard's Frederic Jackson Turner. But the historian or philosopher who fails to fasten this "nuts and bolts" account to the relevant historical and philosophical classics will have missed a chance to experience much of the flavor of this part of frontier life. Harry Mason's time and place is what we might call a fortuitous or auspicious convergence. He captures the flavor that neither Turner nor Marx ever captured. Somehow the boosterism described by Sinclair Lewis in *Main Street* and for that matter, Walter Prescott Webb's thesis about barb wire and six shooters doesn't catch the "essence" described here either. We are assisted more by Wallace Stegner who has helped us define who we were, or thought we were "beyond the 100th meridian" as we set out to build our homes and carry out our lives in one of the last of nature's "gardens." This account by Harry Mason is more than a mere footnote. It is another view, like opening a door on another side of a house for the first time.

We are forever commenting on how fast things change these days. Harry forces us to consider a period of unprecedented change during a time in which machines and fuel made theft of farm fertility possible on the high plains. We have built our nation on this theft, and, unfortunately, we still proudly hail that extractive economy.

Sustainability was on few people's lips in 1920. We were still young and rambunctious, believing that forests, soils, oil and gas would last forever. Fewer people then than now worried about our nation wearing out and about using up what we still call resources; the resources that made it possible for us to define ourselves so heroically. So we have gone from one mining operation to another. One of the best accounts of this history was outlined by the late John Fischer, former editor at *Harpers*, in *From the High Plains*. Fischer characterized the region in which Harry Mason worked as a place that has experienced a series of mining economies. The Indians mining the flint was the most sustainable, for it was an enterprise that did and could have gone on for untold millennia. But then came the white hide hunters and later the meat hunters for the men who built the railroad. Next came the miners of the bison bones to rattle in wagons to meet the freight cars at places like Dodge City for transport to eastern gardens. Then came the miners of grass, followed by the failure of the cattle industry when a combination of drought and blizzard virtually wiped the industry out. Later came the oil and natural gas miners, then the rejuvenation of the soil miners in the teens and twenties in response to a war-ravaged Europe needing food. Finally came the water miners using their Chevy and Ford engines to pump the Ogallala aquifer to raise row crops from Garden City, Kansas, on through the panhandle of Texas.

That's the mining sequence. Not only was Harry Mason in the place of convergence, either through people he saw every day in WaKeeney or through his own direct experience; his life spanned all of the extractive economies on the plains except the one managed by the flint-mining Indians. But even this was almost within reach, as the reader will discover. The buffalo were gone, too, but one local man who had cut and processed their meat was alive and still walking the streets of WaKeeney. He'd started and run the local butcher shop, a natural follow-up once the bison were gone. (He'd also been sheriff for a time.) But now this "retired" butcher had turned the business over to his sons. The old man had come to the area under contract to be a butcher of those monarchs of the Plains for those men laboring to lay the track of the Union Pacific. Now his sons cooled the carcasses of the domestic analog of those wild bovines with ammonia-charged cooling coils and sold their steaks over the counter.

Men who had driven grass-fattened cattle more than three hundred miles to the Kansas City stockyards were alive. Harry repaired many of the tractors of the world's greatest single plowing ever—forty-five million acres of Great Plains grasslands. Relentlessly, often with their lights on late into the night (teams of horses never had headlights), those tractors would plow and disk and harrow, working the ground to a near powder because, it was thought, that is what a good seedbed required. Only poor farmers would do otherwise. After seedbed preparation, these tractors would pull the drills and combines. Relentlessly, Harry would fix them all as they broke or needed an engine overhaul. Later he came home from college on weekends to help pay for that ecological sin, a product of a work ethic virtue. He helped scoop out the fine dirt that was once soil, that near-powder that had blown in under the shingles to accumulate over kitchen and living-room ceilings, which then collapsed, sending plaster, lathe and dust all in a heap to the floor below. He then helped haul this heavy mess to two abandoned dugouts in an abandoned hog pen. Count them: three failures in that one sentence. But there is more. In his memory, the oil and gas industry came and stayed, and the irrigation industry got under way over the Ogallala aquifer.

A year or two before that dust blew under the shingles, college, at nearby Fort Hays State was a convenient escape for him. He liked his psychology professor so, probably for no better reason than that, he studied psychology. It was simply "continuing that line of work," as we Kansans say, that caused him to eventually earn a Ph.D. in experimental psychology. He spent most of his so-called productive adult life as a teacher and researcher.

This book is neither an autobiography nor a biography. Rather, it describes an unfolding drama in which Harry and his father have leading roles. Like all dramatic tragedies, it was destined for a tragic end from the time his father bought his first threshing machine in the 1890s. Both father and son are representatives of the regional character and culture, which is to say that neither father nor son thought of themselves as tragic figures so much as ordinary mortals who, when forced by circumstances, went from one thing to another. Maybe it is because of this attitude that "failure is a parenthetical event" as Mary Beth LaDow says.

Imagine a home on the range where piston rings were being cut from appropriate-sized pipe, where, later, fleets of Model T Ford trucks hauled gravel from washes near creeks when the Union Pacific highway was being raised to a level above the surrounding countryside, and the drive shafts were shortened to accommodate auxiliary transmissions on those same trucks in order to increase the torque, and thereby more effectively haul gravel.

"What did it feel like" to be around the optimistic young owner-drivers, these individual entrepreneurs who were paid by the cubic yard, who would haul gravel for four or five days and then bring the trucks to the garage for Harry to adjust the connecting rods over the weekend?

Harry Mason's history represents a beginning to remove the parentheses from failure. When the garage business failed, the auxiliary enterprises for hanging on required that his parents return to the farm as the final backup. They failed there a second time. This is bad enough, but the real tragedy is that failure seems to go on and on, because we still haven't yet figured out where we are or how to live in a sustainable fashion on our planet.

P R E F A C E

Threshing Machine Canyon

In 1921 or so, when I was a lad approaching puberty on our farm near the small town of WaKeeney, Kansas, my father took notice that he had not taught me much about recreation. He decided to make amends by taking me fishing several times, using bamboo poles from the hardware store and minimum tackle. I enjoyed being with Father but never developed a taste for sitting in the pond-side weeds, slapping mosquitoes and waiting. Father was not an expert, and we caught few fish.

On one of these expeditions we went to a pond in a pasture fed by a very small stream. It was called Silver Lake, and it was reputed to be "no bottom" deep. We fished for hours and caught nothing.

On the way home, however, Father drove us about a mile out of our way, across roadless pasture, to a deep, steep-sided draw. He stopped the car and we got out. Far down among the weeds we could see metal, maybe the remnants of a machine. Father told me that this place was called Threshing Machine Canyon. Sometime in the dim past (though it really couldn't have been that long ago because white people had brought threshing machines to western Kansas only about twenty-five years before), a group of Indians were said to have mobbed a threshing machine and pushed it over the edge of this draw.

Father did not elaborate. He probably didn't know any details; he was not inclined to embroider legends.

But I've thought often of that spot and those Indians and wonder whether they didn't have a lot to do with the stories I'm going to tell in this book, stories about machines coming to a machineless land, about enterprises begun and abandoned when West Central Kansas proved too hard for their pursuers, about (as Wendell Berry says it) people coming to the middle of our continent with vision but without sight. Maybe those Indians saw the threshing machine as a noisy, dangerous monster; maybe the whites should have seen that, too.

Most of these stories have to do with the rise and fall of my family's hopes and fortunes up until I left home in 1934, while prosperity remained stubbornly around the corner. Roosevelt ousted Hoover from the White House, and, in spite of his patrician manner and long cigarette holder, he had sense enough to surround himself with advisers who could start the country turning around. Harry Hopkins, who spearheaded establishment of youth programs and social security, is the man I give credit for creating a climate in higher education where I could get bachelor's, master's and Ph.D. degrees in psychology; notice, and

be noticed, by Isabel Jakway; be married to her in May 1940; sire
four children, two of each sex; and spend the time up to now—
a few months short of fifty years—not free from worry, but
happily free from depression, and very firmly a Democrat. The
happiness has, of course, been considerably deflated. Isabel,
never athletic, became increasingly fragile and short of breath
about 1988, and, after removal of a cancerous bladder tumor in
early March 1990, and early promise of recovery, died on March
28, 1990. It was her quiet suggestion, when I complained of lack
of a project, that inspired me to write the present piece. What she
said was, "Why don't you write something about what it was like
to farm with horses when you were young?"

LIFE
ON THE DRY LINE

THE FARM

Father's
Wheat Threshing
Adventure

As early as July 1897, when my father, Morgan Mason, took his rig out of storage in Delphos, Kansas, to thresh wheat, some farm work was done by engines, though we usually think of this as the period when horses were the source of farm horsepower. Father's rig used a twenty-horsepower steam engine to run the thresher. In an old album, we have a photograph Father commissioned to show his rig in action. There is the steam engine; a long drive belt; a conveyor, called an extension feeder, which is about fifteen feet long; and about six men, standing on stacks of headed wheat, resting on their pitchfork handles while the picture was taken. Their job was to throw the headed wheat onto the feeder. The stacks had been made so that there was just enough room between them for the thresher's feeder, the far end of which connected to the mouth of the thresher, called a separator because it separated the wheat from the straw. When the grain had been threshed and shaken loose from the straw, it was elevated to the top of the machine and poured into a spout that directed it into a wagon that had been made tight and strong so that the wheat would not run out through cracks; the straw went out the back of the separator where it was blown into huge cone-shaped strawstacks that would be used for winter bedding for farm animals or eaten by cattle.

So far, it seems that only men and machines were involved in the threshing, but this was far from true. There would be several wheat wagons, each pulled by a two-horse team, that took turns under the thresher's spout. When a wagon was full, it was driven off to a storage granary or town elevator. The number of wagons needed depended on the distance the wheat had to be hauled. A team could walk at about three miles per hour, and it was often three miles to the town elevator or farm granary, so it might take an hour to get the wagon there; then there was some delay in getting it unloaded. If the weather was not too hot, the team would be permitted to return to the rig at a trot, which shook up the driver enough to make his teeth rattle and cut the return trip to about three-quarters of an hour. The driver was one of the youngest members of the threshing crew. Girls were not permitted to work on threshing crews in 1897; they might have heard naughty words used by the rough threshing hands, and their parents thought this would be a shame. Girls helped in the house, getting meals for the threshers and often flirting with the men when they brought plates of food to the tables.

If a grain wagon did not get back to the rig soon enough, the threshers cleared off a spot of ground where the wagon should have been, put down a piece of canvas and let the wheat run on the canvas. The youth who drove the wagon, who was typically about twelve years old, felt disgraced whenever he didn't get back to the rig in time, but he could not hurry his team too much. He would be scolded if his team was lathered with sweat when he returned. Usually, the youth's conscience did the scolding. Most farmers and their threshing helpers treated the young gently. While a driver's wagon was filling, he could go back to the shade formed by the engine and drink from one of the jugs of water to be found there.

In 1917, when I first did wheat hauling from a thresher, this drinking water was the only water needed in the field; huge gasoline engines had replaced the steam engines of 1897. An important part of Father's crew was the water hauler, who drove a tank wagon pulled by a strong team. This team had to pull the full water wagon up the hills that led down to streams where water was to be had, and steam engines used a great deal of water. The water hauler also had to be strong enough to work the big lever of the pump that brought the water up into the tank. He also had to be a diplomat, because he was taking water which was the drinking supply for some farmer's cattle, and he had to shut gates after him so cattle would not get out. When the creek that supplied boiler water was a bit too far from the stacks of wheat that were being threshed, the water hauler was under pressure. About all he could do was to pump fast. If he was late, he would not only be criticized, he might be fired from his job. The engineer who ran the threshing engine could give him a bad time.

The engine had a glass tube attached to the back of the boiler, and in this the engineer could see how much water there was in the boiler. When the water level fell beyond a certain point, the engineer knew that a soft plug, made so that it would melt before the boiler got too hot, would blow out. This would disgrace the engineer and might cost him his job. When he saw the water level going down, the engineer would blow the whistle, which could be heard by the water hauler, but he didn't like to do this, because the whistle used up steam and thus caused the water level to drop even faster. Usually, the engineer would make a last mournful toot on the whistle and shut the rig down until water came. If he let a soft plug melt, he would lose steam, water and heat and would have to start over again, just as if it were four-thirty in the morning. Threshing would be over for whichever half-day was in progress when the plug blew. If the blowout occurred in the forenoon the engineer missed his dinner. (The noon meal on a farm is called dinner because it is the heaviest meal of the day.) After he had screwed in and tightened a new soft plug (he was supposed to have a spare or two), he would kindle a fire with straw, build it up with sticks of wood and finally go to the nut-sized coal called steam coal. He would watch the gauge on the steam dome creep up toward 150 pounds per square inch, which was working pressure. If he got diverted from watching the gauge, the pressure might build up as high as 180 pounds per square inch, where the safety valve would open. This didn't cause an emergency, it just wasted a bit of coal, water and steam. The engineer might even sound the whistle to call the crew back to work, thus easing the boiler pressure a bit more.

What might have diverted the engineer from watching the gauge? It might have been dream or reality, but it would likely have been concerned with the farmer's older daughter bringing a bit of food and an opportunity to flirt to the best husband material the threshing crew could offer.

F I N A N C E

My father probably thought of his work threshing wheat in central Kansas in the last years of the last century as an adventure. It might better have been called an escape from his family. He said enough from time to time to suggest that his own father, whom he called Pap, was always in debt, and that any money he earned while living at home would go to pay interest on debts, to pay for medicines for rheumatism or for music lessons for Father's older brother. So it seems that Father would help Pap and his older brother do the spring plowing, corn planting and cultivating, then with the corn "laid by," he would ship his team by rail through Kansas City and on to Concordia, Kansas, a straight line distance of nearly two hundred miles, but probably nearly three hundred by rail.

In each of those years, the first man Father would take onto his payroll in July was Wally Warren, his engineer, who would draw five dollars a day until the rig was put back in its shed in Concordia, late in September or early in October. The next man hired would be Morgan's separator tender; next would be a man to handle the water wagon and drive Morgan's valued team, itself a trust not to be taken lightly. This "water monkey" would also lubricate relations with farmers whose streams and ponds would furnish water. The water hauler drew three dollars a day.

At the top of the clean-grain elevator on the thresher sat a grain weigher, set to accumulate fifteen pounds. When fifteen pounds of wheat had accumulated, the weigher dumped it through a spout into the grain-tight wagon. A counter, set at zero at the beginning of a job, and counting one bushel for every fourth dump, told how much the machine earned each day, at five cents per bushel. Each night at quitting time, the separator man would write down the reading in a little book he carried in a shirt pocket. On a good day, this could be eight hundred bushels, earning forty dollars. The farmer wasn't expected to pay until the job was done, unless he was regarded as a poor credit risk. But Morgan early learned to weed out the skinflints, to earn the custom of the established farmers and the young farmers who would, they hoped, soon be part of the establishment.

Men to haul bundled and shocked wheat to the rig or to pitch headed wheat into the thresher from stacks were paid as much as a dollar a day by the farmer whose grain was being threshed. The farmers also provided food; farmers' wives and their big daughters would gang up in help-one-another groups to feed threshers, and each wife, and husband as well, wanted to meet the expectations of the community. If the food was good, the men would defend the women's reputations as cooks by eating everything set before them, and they'd smile at the farm girls as they served the food, and maybe give their buttocks a sly pinch if the girls brushed close and appeared to welcome it.

I doubt Father did much pinching, for he felt loyalty to his fiance Etta, back in Iowa, and to his own ideas; a proper young man was likely to become rich more quickly than a frolicsome one. With a third-grade education, Morgan was not likely ever to become a lawyer or doctor. He had no use for preachers, including his brother Charley, and he thought of being a farmer only as a necessary apprenticeship. Father had all the skills needed for farming, but he could not put his heart into it. Somehow, he would become involved in finance. He was good at figures. It probably seemed to Father that, with pluck and true grit, everything was possible.

Bundle threshing would be Morgan's first order of business, and his first source of cash each year. Smaller farmers would often bind the slightly green wheat to spread the harvest season out over weeks, rather than days, using their own labor instead of itinerants who would come to the area as wheat heading, rather than binding, got under way.

Before the season got rolling, Morgan's crew would need advance wages, his team would need feed and Morgan would need to buy coal to raise steam. Each year, Father would open an account with a one-thousand-dollar balance written into the stub of the first check in a checkbook this banker gave him. One thousand dollars was the face value of a note secured only by Morgan's good name. Each October, he would close the account, his closing balance draft being debited by the thousand plus a few dollars of three percent interest.

Morgan would spend little time at the rig during threshing days. At four-thirty in the morning, Warren would be at the rig, shaking clinker out of the grate and starting a new fire with a handful of straw, then more straw, then small nut steam coal, added a bit at a time. Morgan would appear before six o'clock with whatever breakfast the threshing hostess provided for the engineer, carried in a dinner pail or saddlebag. By six o'clock the water man would have started for his first load of water, the separator man and pitchers would have arrived, the glass tube on the boiler would show at the high mark, and the steam gauge would show 150 pounds per square inch. Threshing would begin. Morgan, on his hired saddle horse, would be long gone.

First, Morgan would catch up with the water wagon to see that his team was being well treated, that the pump was not drawing dirty water, that the farmer who was furnishing the water was happy with the way the water wagon was treating his pasture and that the water monkey was closing the farmer's gates. Then, if the rig was still threshing bundles, Morgan would check with a farmer whose wheat was being headed and stacked to agree on the date when the rig, finished with bundle threshing, would pull up to his stacks. Then he would ride across the prairie to the next job site, scouting the way for Warren to guide the clumsy-wheeled engine, that pulled a separator, a feeder and the junk wagon that contained a bit of everything and that would be left sitting near the engine. It was Warren's wagon, and he had painted his initials, W.T.W., on its sides with a broad brush.

When threshing from stacks, the machine had to be pulled between them and would be faced so that the wind would blow from the extension feeder toward the straw stacker. When threshing bundles that were hauled to the rig by hayracks hired by the farmer, however, the rig would be set up where bundle haulers would drive short distances to the rig or where the farmer wanted his strawstack to be.

In siting the machine, shallow trenches would be made for the front wheels of the separator so that, when Warren eased the machine back to its working position, the separator would be level from side to side, with its rear wheels slightly higher than the front; straw and chaff would be blown uphill. This kept the several fans and the wind stacker from hurrying the straw through the machine and tended to calm farmers who were suspicious that rapid threshing would put too much grain into the strawstack. Farmers had reason, however, to hope that the threshing would go rapidly. Feeding a crew of twelve to fourteen men well enough to maintain his family's reputation was expensive, and every day saved was a bonus. Kitchen emergencies were luxuries a farmer would well do without, and itinerant laborers would not stick around idle without pay. It is little wonder that the industrial workers of the world eventually found threshing crews good prospects for organizing. At about the time, 1917 to 1918, when such efforts might have succeeded, World War I interfered. Any effort to organize farm labor would have been squelched as unpatriotic.

AFTER
THE FIRST FROST

Father threshed in North Central Kansas yearly for at least three years, probably more. Not all the years were monotonously the same. One variation is suggested by the presence in a desk drawer, during my youth, of a sepia-toned studio photograph of a young woman, obviously conscious of her charm, bare-shouldered and with a blouse that more than hinted at an ample cleavage. I do not remember what induced Father to "explain" the picture to me, but I remember my disappointment. He said he had just wanted to show a young farmer that he could steal his girl. I felt nothing for the girl, who had penned a tender dedication to "Morgan," my father, but for the first time I saw Father as a ruthless competitor. That was the accepted pattern for a young man at the time and later—I should not have been surprised—but I was not ready to cast aside childhood and see my sire as a man, not a god.

More acceptable to my youthful ideas was the story of a weekend emergency with the rig. Part of the profitability of threshing had to do with spending as little time as possible getting from one job to another, and the broad-cleated drive wheels of the steam engine were needed to make their own road for the separator and feeder trailing behind. To cross creeks or rivers, the crew had to keep planks at hand to reinforce floors of the bridges, since the engine and separator were heavier than the intended loads. Apparently, on a Friday or Saturday afternoon in 1895, plankings gave way on a bridge over a creek mouth on the Republican River. The rig was dumped into deep water. Father did not hesitate. He caught the evening train himself, got the dealer down to his store, signed insurance papers, signed new notes and superintended loading a new rig—presumably engine, separator and feeder—on railroad flatcars. Then he accompanied the machinery to Concordia, Kansas, presumably as a caboose passenger. He was in the field, he said, ready to thresh on Monday morning. Our whole country was younger then. I'm sure such a mobilization of technology would not be available to a twenty-odd-year-old entrepreneur today. Bureaucracy eventually fouls commerce, even in capitalist countries, in spite of computer technology.

When the threshing season was over and the rig had been put into safe storage, Father was free to head home to Bedford, Iowa, near the state corners of Iowa, Missouri, Nebraska and Kansas. He usually did not hurry, but spent several weeks husking corn in the fields of Northeast Kansas. He had his team—the saddle horse shown in the picture of the threshing rig must have been a livery barn rental—and he could get feed for the team, board for himself and one cent a bushel for husking corn. Father's husking peg, with leather loops for fingers, hung in what we called the buggy shed on the home farm near WaKeeney. It was the anchor for Father's corn husking story during my youth. He could husk a hundred bushels of corn in a day, after a week or so of conditioning, and thus count on a dollar a day pure profit. He needed the money. As he told it, his own father had an insatiable appetite for money, to pay interest or to buy treatments and medicine for himself and Father's older brother, Charley, or to pay for Charley's lessons on the parlor organ. Musical skill was part of the life Charley was preparing for—that of an itinerant preacher. I gathered that neither Pap nor Charley was very sick, nor very well either. Such an appraisal of farmers' health seemed to be quite common, even during my youth. It was probably part of the folklore of a segment of society which held fast to Herbert Spencer's dream of progress in the face of the growing certainty that ordinary lives, full of hard work and a fat-rich, vegetable-poor diet were to be the farmer's lot.

In 1898 threshing season was especially different, and a near-disaster for Morgan. All the machinery performed well, but the weather was against him all summer long. Rain ruined the last of the threshing; some weeks the weather was so wet that the crew did not go to the field on more than two or three days. Wheat began to mold and rot in the stacks. Finally it sprouted, so that any expectation of salable grain was quenched. By the middle of September, there was nothing to do but to put the machine into shape for next season; drive it into the rented shed; drain the boiler; pay Warren, the separator man and the water monkey; and book a railroad car for his team to Kansas City. There Father would visit the threshing machine dealer, pay accumulated interest on his loan, renew insurance on the rig, put down what he could on the principal of the note and reschedule the freight car carrying his team for Creston or Bedford, Iowa, the two railroad stations nearest the home farm.

But Father did not return speedily to his family home that year. I am not sure whether a spell of cornhusking followed the threshing. Probably not, because he joined a cattle drive, starting at the stockyards in Kansas City, Kansas, and ending at the Harmon Ranch on the Smoky Hill River, twenty miles south of WaKeeney, Kansas. A man named Roy Dunnington was the trail boss. Probably he met Father at the Kansas City railyards. He needed a man and team to drive the cook wagon and cook the meals, and Father and his team filled the bill.

The story came out in another one of Father's explanations. As I remember them, these were stories Father liked to tell, not ones dragged out of him by a teasing child. This one explained the nickel-plated five-shot revolver that hung from a nail, high on the wall of the buggy shed, near the nail that suspended the husking peg. According to Father, anyone associated with a cattle drive was not decently clothed without a Stetson hat and a revolver, though there was no holster for Father's gun, and presumably none was needed since the cook used to sit on a spring seat on the cook wagon instead of riding astride a cow pony. (It is very unlikely that Father's Stetson hat was as broad and high as those you might see in Western movies. Throughout my youth, I remember that he wore a tan hat much of the time.)

I have no idea of the size of the herd driven to the Harmon Ranch. It might have been anywhere from less than a hundred head to two or three hundred. I also know little of the time required to move the herd the 350 miles from Kansas City to the ranch. At seven miles a day, which was considered a day's travel for an ox team on the Oregon Trail, it would have taken about fifty days, which would have brought them to their destination about mid-December, perhaps as early as late November.

The Kansas City stockyards are on the Kansas side of the Missouri River. Prodded to their feet, urged to drink and led by an aggressive cow, the cattle would have strung out up the slight hill to the northwest. Within a mile they would emerge onto a broad flat of prairie, a gentle slope leading west and downward, covered with grass and with a few wagon ruts, the older and deeper ones already reclaimed by grass. The lead cow, raw boned and lacking a calf to hinder her, took charge, and other, lesser cattle followed. Two experienced cowhands would ride ten to twenty yards wide and a little behind the lead cow, moving in to push stragglers back to the main bunch or to direct the herd, taking it easy, occasionally calling out in what they hoped were soothing voices. Two more men on horses followed, keeping abreast of the wide outline of the herd; behind, to deal with the frisky and ever-curious calves, four less experienced men rode, trying to absorb Dunnington's instruction that the calves would probably turn toward their mothers once they found themselves alone, and the main thing was to let the herd drive itself, intervening only when an individualist looked lost. At a glance, an observer would think that the point and swing riders were the amateurs and the rear guards were the professional cow hands. Certainly the forward contingent had the older, more tacky clothes; they slouched and their saddles were worn and scuffed. The younger, straighter men at the rear were in shiny saddles. They wore leather chaps over their Levis, and their boots were newer and less scuffed. But the horses they were riding gave them away. The older men rode the better, taller, often younger horses that knew what to do and not to do too much. Dunnington, a big man, would ride a bay gelding of about a thousand pounds—eleven hundred when fat from a winter's rest. He kept about twenty yards back, and to his right and a horse's length back came Morgan's Tom and Jim, pulling the cook wagon, with Morgan in a spring seat ahead of the boxy body of the wagon. The cook wagon resembled a scaled-down caboose, without the cupola. It contained a topsy stove with four lids, suitable for burning wood, soft coal or cow or buffalo chips. Morgan would splurge by using coal for the first noon meal; wood would be gathered by the creeks up toward Tonganoxie, and suitably dry chips would be picked up and carried along in the wood box whenever an old bison bedding ground or wallow provided a concentration of them. Morgan would go ahead alone after the noon meal, proceeding while the cattle rested, setting himself up for supper at the night's camp spot. His team could graze while the herd was catching up and after supper. The herd would keep near the course of the Kansas Pacific Railroad, but far enough into the prairie to be out of sight of towns. For the most part, the herd stayed away from established farms. Tonganoxie, Lawrence and Topeka would pass, then the herd would bear slightly north of west, curving back to the Smoky Hill River to pass Junction City, Abilene and Salina. Then, leaving the river to bear a bit north, to avoid a deep river bend, the herd would follow the Saline Valley, bending back to cross under the railroad near Dorrance, then back to the Smoky Hill, nearly twenty miles south of Hays, and on to the Harmon Ranch.

All that can be known surely is that Morgan fixed WaKeeney in his mind as a possible place to settle down should things not work out in the Concordia area where, with his wife

Etta, he would settle in 1901. It is possible that the herd followed the railroad more closely west of Dorrance, instead of staying near the Smoky Hill. Father once boasted that, in 1898, he had roped a calf in the courthouse park in WaKeeney, presumably while passing through the town. Such a route would have allowed the herd to graze for a day or two in the broad valley of Big Creek, then spend a couple of days cresting the rise of ground to descend again, on a gentle slope, to the Smoky Hill River and the ranch.

I know that the herd passed near Dorrance, Kansas, because Dunnington was short a man, and he recruited Fred Spaulding there. Spaulding was already established at Dorrance in '98, and business sometimes brought him to WaKeeney during my boyhood. He was always invited to our house for a meal, and my mother knew that he would expect a piece of cherry pie. I remember him only as very tall; he stooped to get through an ordinary doorway, and I think I remember that he placed his modest-brimmed Stetson beside the pendulum clock on our clock shelf. I've often wondered how good a camp cook my father was—he once roasted a Christmas goose to dry slivers on our farm stove after we had moved to town and Mother trusted him to do this chore—but Spaulding, at least, survived the drive, and he was a cheerful guest at dinner; Mother did the cooking on those occasions. Spaulding remained at Dorrance during my youth. Father received a yearly invitation to a watermelon feed at the Spaulding ranch, which would have occurred in late summer. I think the Spaulding ranch also grew cantaloupes for market.

I remember only one story of camp life on the cattle drive. Dunnington seemed to have a good-humored contempt for the marksmanship of his cowhands, and Father told me of an evening when they set up empty tin cans as targets and practiced with their revolvers. According to Father, anything the size of a gallon can would have been quite safe on the firing range. On the occasion Father mentioned, the cowhands had emptied their pistols when a jackrabbit hopped across the prairie and sat up about thirty yards from the camp. Father, who had not been shooting, took up the nickel-plated .38 and, pointing it like a finger, killed the rabbit dead with a single shot. Dunnington's comment was, "Better put it away, Morgan, there's no point in ruining a good record."

How did Father get back to Bedford, Iowa, from Ransom, a bit south of the Harmon Ranch, or from WaKeeney? The Union Pacific's road from Kansas City to Denver was completed in 1870. I presume Father drove to WaKeeney, put his team on a freight car, and rode the caboose to Kansas City, where he could make a change of railroads and get to St. Joseph, Missouri, and probably on to Bedford, or some near point. I'm sure he went back, because on February 6, 1901, he married my mother at Bedford, presumably at the Methodist Parsonage.

A Young
Couple Settles Down

My mother once told me that she first "noticed" Father when he came to the Putnam farm near Bedford, Iowa, to return a farm machine his father had borrowed. She was impressed by his skill and gentleness with his team as he maneuvered the wagon to unload the piece.

Mother's family seems to have been well respected in Taylor County, Iowa, farming land her father had bought after being mustered out of the Union Army, disabled in training camp with "consumption," which was what tuberculosis was called by common folk. Mother had one older sister, Flora, several brothers, and a much younger sister, Stella. By 1901 it is likely that brother Jason, oldest child of Mother's parents, Henry and Cornelia Putnam, was in charge of the family farm, and that Flora, who had married John Fitch, farmed nearby in Taylor County. My impression is that the Putnam and Fitch establishments were well-equipped and respectable, with ample barns and purebred cattle—the sorts of places where lesser farmers could take their cows or mares to be bred, or where they could borrow a cultivator.

Two questions arise when I attempt to place Morgan and Etta in context in Bedford. Why did they not settle near their parents? Why did they find themselves twenty-nine and twenty-seven years of age at marriage, early in 1901? There is no way to get into their skins or their diaries to throw light on these questions, but there are data.

In an autobiographical sketch handwritten when she was seventy-nine, Mother writes that she spent a year at Normal School in Stanbury, Missouri, qualifying for a teaching certificate. She once confided to me that she did not, in fact, stay there for a full term. She said that she had been miserable with homesickness, weeping uncontrollably, and that she had come home. The fifteen years between returning from Stanbury, which I estimate to have occurred in 1885, and her marriage in February, 1901, were taken up, according to her manuscript, by teaching in country schools and by work in the office of the Taylor County, Iowa, Superintendent of Schools. During this time, she was available as a "supply" teacher, that is, a substitute for teachers temporarily disabled.

I remember Mother as a homebody during my boyhood; she was no gadabout. She preferred to stay at home on Saturday afternoons instead of shopping in town and gossiping. Her groceries were ordered by telephone and picked up by Morgan on his way home. When I was about twelve, neighborhood ladies formed what they called the Ladies Country Club. (It was renamed Country Ladies Club sometime in the 1920s, when some of the members saw that the earlier name signified a much more free-thinking posture than they intended.) This club was the only purely social activity Mother had. Morgan would not go to church, and she would not go without him, but I do not think she missed the church life a lot. In short, I think the shyness that drove Mother home from Normal School in 1885 persisted throughout her life and that what might have been the usual courting and mating years—in that culture from sixteen to twenty—passed without enough social activity to encourage beaux. In 1900, it was now or never! A lonely occupation and spinsterhood loomed; it is no wonder that she watched Morgan through the kitchen window.

What about Morgan? Why was he unwed at twenty-seven? Here the answer seems to be that his family was not healthy—some or all of the four members seemed to be too sick to carry out normal duties much of the time, and Morgan, as the younger son and the lowest in authority, drew the duty of doing the farm work, whether he felt well or not. He did not talk much about his family, but I remember anecdotes in which he walked behind a cultivator, its plow handles holding him up because he was too sick to walk steadily without their support. He once related to me how he came to see himself as the mainstay of the family, a better man than his older brother Charley. In times past, their father had sent them always to the same field to husk corn, and Morgan could not bring himself to compete. He was content to keep his wagon abreast of that of Charley. On one occasion the boys were sent to separate fields, out of sight of each other, and Morgan, unloading his wagon at the crib, asked his father, "Where's Charley?" "Charley ain't in yet," his father said, and Morgan suddenly realized, he told me many years later, that he was the better man. What the family needed, Morgan would have to provide. During the weeks, he would fall into bed soon after supper; family mores, as well as exhaustion, prevented socializing at Saturday night dances. Charley's loud and fervent praying turned him away from attending church. There was little chance to socialize or court.

Why did Morgan and Etta not settle near their parents, in Taylor County, Iowa? Morgan's personal history suggests that he might have been eager to leave the home scene behind and escape the burden of being the least in the family hierarchy, the only member who could not afford to be less than able-bodied, no matter how sick or healthy he felt. Etta would go where her man led, as was the custom in those days. The newly completed railroads and the young towns of the Kansas prairie region advertised for settlers in all the Plains states. Railroads, and land companies as their agents, scoured European countries where immunities from military service, granted to the progeny of invited farm settlers, were running out. The temper of the times encouraged young farmers to move west. What they hoped for was affordable land which, tended according to the practices of Iowa or Europe, would yield abundantly. Like everyone who is not born rich, they hoped their children would be spared the privations and hard labor of their own youth. The possibility that they might someday be rich but unhappy probably never entered consciousness. First, they would work hard to become rich.

Settlers recruited by land companies were directed to specific communities. Morgan was on his own and so needed to have picked a landing spot. He had seen Concordia and WaKeeney. There they would go.

But before we see them in Kansas, there is the matter of religion.

Mother was a member of the Methodist church, but Father missed out on that honor, probably with little regret. As he or Mother told it, he was proposed for membership and his name was posted on a list. There was a probationary period, during which any member might object if he thought the young man unworthy of salvation through Methodism. During the probationary period, Father was scheduled to go to Kansas to thresh wheat; an appeal that the probation be shortened was denied. He did not renew his application. Father was happy to let all of his children attend Sunday school and even join the church, but he held an aloof, superior attitude toward the clergy, keeping a respectful distance from religion during his life. Mother would have been free to attend church, but she would not ask Morgan to harness a horse to the buggy and take her two and a half miles to church. Father's brittle attitude toward organized religion troubled me during my youth. Emotional, undersized and given to fantasy, I was an early candidate for salvation. Only in my own later years have I seen his position as the beginning of my own agnosticism.

Etta Putnam
Mason Remembers

Mother left an account of Morgan's and her first days together. It follows, though I've edited it for clarity from a handwritten manuscript made in 1961 when she was eighty-nine years old.

In February 1901 [I] married Morgan W. Mason, a farmer and auctioneer in Bedford, Iowa. We came to Republic County, Kansas, in March [to a farm] located fourteen miles from Concordia, Kansas. Our wedding trip was by train, starting early in the morning. Our first stop was St. Joseph, Missouri, about noon. We were hungry. [We] ate our wonderful lunch—which my mother had fixed for us—of fried chicken, Iowa fruit, pickles, cake and cookies and sandwiches in the waiting room of the depot. I did not miss hot coffee as did my husband. Those were the days before Thermos bottles. We were Scotch, using the plain way. Water was our drink. We enjoyed our lunch and

the resting place in the depot at St. Joe; the next
train ride was longer. When we arrived at Wayne,
Kansas, in the evening, we were tired. [Wayne was]
a very small place a mile from where we were
going to live. [We] thought to rest until the morning, going up an outside stairway to our room. We
slept very well and ate breakfast there. It was a
beautiful March morning. We enjoyed our walk of
a mile. The first thing we saw was Morgan's team
of fine horses in the barn and our young dog that
met us. The house, which was a large one, [was]
mostly on the ground floor. It was built of three
different materials: cement, stone and wood. This
home-furnished room was in [a] separate room,
and Morgan's parents and brother lived in the
other part. The owner of the place was quite old
but active; [he] boarded with Morgan's folks. Being
a hobbyist, he had gathered much of interest about
the place. With his big mule team and strong
wagon he would spend much time gathering all
kinds of ston[e]s large and small, bring them home
and make use of them on the place.

In a different account, Mother wrote:

> Our part [of the house] consisted of a large
> bedroom and living room in which we placed our
> furniture that we bought for fifty dollars at a
> secondhand store in Concordia, Kansas. It con-

sisted of a cookstove, a table, four chairs, a bed, a
washstand and bucket, a washboard and tub,
culinary articles and some much needed dishes.

We enjoyed the place; it was much like
Iowa. The man we rented from was sort of a
hermit; I don't know whether he was a widower or
bachelor. I never did ask. He lived happy at home
with his large mule team and strong wagon. He
made daily trips gathering all kinds of fruit trees
which he planted and cared for in a bachelor way,
very careful. He brought a large flat white stone,
four feet by three feet for my doorstep. I used to
enjoy keeping it white and clean. Many trees were
bearing fruit at that time. Mother Mason and I
canned plums and peaches that year for next year.
Over a small stream the landlord made a bridge of
stones. It was an artist's piece of work. This stream
was pebbled bottomed. The train came by every
morning quite early; it made it seem like home. I
never was homesick for Iowa.

We probably would have stayed there, but
the owner's fields were all hedge fenced as were the
cow lots and pig pens. Morgan planted corn, wheat
and oats. Sixteen feet around fields the hedge roots
sapped the moisture to the extent that nothing
would grow for that distance; we decided to get us
a home, which we located earlier, loading our
cows, horses, furniture, some chickens and turkeys
into an emigrant car. Morgan and his father came

to Trego County where [Morgan's father] had
bought 160 acres of land three and a half miles
from WaKeeney, Kansas. We rented a place from
George Baker, who lived in town. He was a butcher
who furnished meat for railroad men working on
the Union Pacific Railroad to Denver, which was
completed to Denver in 1870. While the men took
our goods to WaKeeney, Mother [Mason], Charley
and I stayed with people in Concordia. Our men
unloaded everything, put the stock in the corral
and the furniture in the house and lit a stove to
warm us. We all lived there until Morgan and I
completed our sod house on our own place and his
folk still lived in the rented house. We moved into
the sod house and lived there two winters and two
summers. We had completed the house except for
shingling the roof when one morning, as Morgan
and I were going over to our place to work [I drove
the horses and Morgan held the plow to cut sod],
we met a thresher man who wanted an engineer
for two months and would pay five dollars a day
until he was finished threshing. We hated to leave
our roof that long for fear of storms. We talked it
over; money seemed good then so we decided for
Morgan to be his engineer. During the time we
didn't have any storms. We could finish our soddy.
There was lots of straw in the country and Morgan
smoothed the floor and packed straw for the floor.
He built two-inch lumber around the sides and

ends of twelve feet by thirty feet, the size of the
house and staked it solid. Mother Mason had given
us an ingrain carpet she had, so we fastened the
carpet to the two-inch staked lumber. I don't know
how the packed straw made such a solid floor, but
it did, as nice as you ever walked over. In the
middle of our house two telephone poles held the
slanting roof; this made a division for two rooms,
living room and kitchen, around our stove and
eating table. I bought a table oilcloth and sewed it
to the carpet. That way we kept the carpet clean by
washing the oilcloth when needed. Our roof
slanted from west to east, where our windows
were; as the east side was dug below [ground] with
windows above [ground] we had plenty of light.
Our bed was on the west side sloping to east—that
part of the room was lower. We got funny papers
and pasted them over our bed. They were lots of
fun to read. Our bed was dressed in white with
valance to the floor. Morgan made us a wardrobe
for our clothes which was curtained with the same
material as our doorway between rooms. I brought
my sewing machine [Singer] and my organ with
me. When I married I had saved one thousand
dollars from teaching etc. My father Putnam gave
me two hundred dollars, the price of a cow. We
had bought some cows and, with Morgan's team
and wagon and our furniture, loaded them into a
car [the emigrant car mentioned earlier] and came

West. We had a man who witched for wells find us
a wonderful well. Roads were in the most convenient places. Then [later] grass land was being
plowed. Several good crops were realized, so others
[ranchers] followed suit and cattle, their main
dependence, were being sold. When this happened, cattle sales were good and they brought
cash. One February, Morgan had a [farm] sale every
day but on Sundays. He drove a nice black team to
sales in two counties, Trego and Gove. We farmed
some and kept stock, kept a hired man to take care
of things and be with I and the children when he
would be late getting home. In 1904 we built a
frame house near the well, pumped water with a
windmill. Mary was born January 14, 1905, in the
new house. The hotel in WaKeeney was changing
its heating plan and Morgan bought their large
hard coal base burner, which, with isinglass gave
light and heat for our two rooms and kitchen. Any
time in the night I could get up and take care of
Mary, without any lamp light, and [it] was always
warm—a great blessing for us for many years until
we could not get hard coal anymore. [We] used it
for all the children; Howard was just a year old in
1919, when the top of our big barn was burned.
The base burner [stored in the barn] fell into the
basement and broke. The lower part of the barn
was of made of cement, so we rebuilt it and it still
stands in 1961.

Mother's manuscript ends a few sentences farther down the page. She tells how, in building the frame house, "Morgan blasted rock from the side of a hill across from where he wanted to build. With this rock and cement he made the foundation for our house; it had a bedroom with a closet and a dining room. [For a] kitchen we moved a building that had been used as a granary, [connected] it to the other part and built porches on the east and west sides that were screened to keep the children in."

By the time when I can remember the house, the screens were no longer needed to contain the small fry, but were useful in excluding flies, which swarmed around the house and barns, were trapped in screen-wire traps, and immobilized on Tanglefoot Fly Paper, which came in castor-oil-glue sheets, stuck together until needed for use. When ripped apart, they sat on tables and other surfaces, sometimes with small saucers of vinegar, sugar or both together, to serve as fly bait. Horse farming apparently made farm premises attractive to flies. They were never so bothersome in town.

LAND AND PRESTIGE

I can attest that Mother's 1961 account of her life dealt in sanitary ambiguities. I heard bitter emotional soliloquies from her during rainy days in her kitchen while I built corncob corrals and entertained my own fantasy as I played on the floor near the stove. The thousand dollars she had saved before marriage had been frittered away; the milk cow that was supposed to have been bought with the two-hundred-dollar dowry did not materialize—it was lost somehow, blended in the interminable buyings and sellings that characterized my father's style of livestock management. Mother's written autobiography lacked a forthright emotional tone. History had been rendered to fit Victorian propriety.

What circumstances in Mother's account might have justified disappointment, without assigning personal blame to her husband? First, she had married into a family of four other people in 1901, and she was not rid of them, day or night, until the summer of 1904. Today such an arrangement would almost automatically lead to divorce; not then, and not for my conscience-driven mother. (This family bitterness endured, as I recall, and wasn't confined to Mother. During my childhood, Charley visited us several times. He held fervent prayer sessions with Mother and exhorted Father to be saved. After a day or two Father would "lend" Charley fifty dollars and bid him goodbye. Charley eventually settled near LaJunta, Colorado, married, and raised two sons, though he was a bachelor until middle age.)

Not visible to the farmers who moved west with the encouragement of land speculators and the railroad, rainfall tapered off dramatically as one moved West. The 1938 *Yearbook of the Department of Agriculture* gives rainfall data for Bedford, Iowa, averaged back twenty-two years from 1938, at thirty-two inches. For the forty years before 1938, the comparable figure for Concordia, Kansas, was twenty-five inches; for WaKeeney, where most of Mother's married life was spent, the average was twenty-one. It was not necessary to blame the Osage orange hedges for "sapping the moisture for sixteen feet" in order to account for slower growth of crops at Concordia, though the hedges made farming more difficult; teams had to be turned more shortly; at ends of fields because of them, and undoubtedly they took their toll of fertility to support seldom-rotated crops. In Trego County, wider fluctuations in rainfall were added to the handicap inherent in smaller totals. Father had bought land right on the dry line (the 100th meridian).

These facts were masked, possibly, by the eloquence of real estate promoters, who had bought the land originally granted to the Union Pacific Railroad near its right-of-way. I have heard different accounts of the extent of this strip—some say ten miles on each side of the track, some say five. It seems likely that railroad lawyers and land agents exercised judgment and were given power to substitute better, more level land for poor, hilly sections. My father used to delight in an account of one of the early real estate agents, doubtless a brother-under-the-skin of the man who sold him a rough quarter section two miles from the center of WaKeeney for about seven dollars an acre. (His own father, I think, paid a dollar an acre less and got somewhat better land, about two miles farther from the county seat.) The real estate man, said Father, would drive buggy and prospective buyer across the prairie, pause by a rain-filled buffalo wallow, take out his tin cup and swig water from the pond. Then he would smack his lips and say, "I haven't seen this spring so low in ten years."

Father was dedicated to progress—his own and that of the community as a by-product. Father seemed always to have been a Republican Party County Committeeman. He bathed in a washtub near the kitchen range and shaved carefully in preparation for sale-crying trips with his team, and later in a Regal 30 horsepower car, still later in Model T Fords, sold off and replaced yearly when business was good. During World War I, Father was engaged almost every Saturday in an auction of donated items to raise money for the Red Cross or in promoting the current Liberty Bond drive. As a public figure associated with horseflesh, he started the trotting and pacing races at Trego County's yearly fair, billed there as Colonel Mason—a title country auctioneers seldom or never claimed, but encouraged others to use.

In spite of this, he never quite qualified as one of the boys among the powers of the community. I do not think malice kept him on the fringe. There is an historical background of which he could not have been more than dimly aware.

It may not be fair, but prestige granted is usually more secure than prestige earned. In retrospect I see that earlier settlers—from soon after the Civil War—who had English or

Scottish surnames, were WaKeeney's upper crust. Some of these were doubtless remittance men of old English families—black sheep—paid to stay in America and thus out from underfoot. Another circumstance granting prestige was to have home-steaded land before the railroad came through. This gave advantage in experience as well as in family acceptance. Some of the earlier settlers in Trego County had raised families and sent their children to Kansas University at Lawrence before 1904. I remember that once my mother spoke to Mrs. Charles Hille, the wife of one of WaKeeney's bankers, telling of her amusement with the happy gaiety of the young who were returning from the university on the train. Mrs. Hille, who would have been Faith Wonner at the time referred to, said with shy embarrassment that she had doubtless been utterly silly and foolish, whereupon Mother reassured her—quite unnecessarily. I felt confused when I heard this exchange as a youth; now I see it as evidence that once prestige is granted, one can afford to be spontaneous. Mother couldn't, and Father overdid it. Success as an auctioneer, being a trusted judge of livestock, the fellowship of the Masonic Lodge and the Mystic Shrine, community service in country school box-supper auctions and World War I fund raising were more or less the waste motion of fading youth.

THE AUCTIONEER

Our bookcase on the farm south of WaKeeney held three or four heavy books that no one ever read. (The Bible was not neglected. Mother read this often and I remember reading most of it during my adolescence.) One of the neglected volumes was *Morals and Dogma* which seemed to be an explanation of the Masonic order. Father belonged, as did most WaKeeney business-men. Second on the unread list was *Bigelow on Torts*, bound to emphasize that it was a law book. I tried it and found it dull. Mother confided to me that Father had once started home study in law but gave it up. This prepared me for what I was to learn later when advising veterans of World War II. Few students ever go beyond the introductory assignments of correspondence courses. Their reasons were probably the same as Father's. His first years of marriage were taken up with breaking sod and herding his own and other people's cattle on the buffalo grass that paved the landscape of Western Kansas. Children arrived in 1905, 1906 and 1908. In 1918 brother Howard was born. By that time Father's dream of being an educated man must have faded, though he continued to hope for conspicuous success as a businessman until the dust storm and Depression years of the 1930s.

The bookcase held two books that bore marks of use. These were *The Auctioneer's Jokebook* and *How to Become an Auctioneer*. Mother once explained that Father could not go to school to learn a profession, and that he saw that our county and adjoining ones did not have an auctioneer. Furthermore, one could become an auctioneer by showing that he could do the work. Father had stationery printed which showed a postage-stamp portrait in the upper left corner, advertised in the *Western Kansas World* and was in business. He charged two percent of sale receipts, and, in his most profitable period, just after World War I, this translated into seventy-five to one-hundred-fifty dollars for a day's work. Many sales were of the effects of farmers who had given up Trego County, and who had to clear obligations before moving on.

Father was a good judge of livestock and worked hard for the highest price he could get. When an animal or a piece of machinery lacked for bidders, Father often bought it. The auctioneer did not bid openly, of course. There were always trusted friends present, and they were never surprised when something was knocked down to them. Bank officers were usually the sale clerks. Through them good citizens could arrange credit, even at cash sales, and having their man on the scene protected the bankers against slippage; furthermore, they took out one percent for the clerk's labor. Father was encouraged in his auction career by local bankers. Other people finally advertised to cry sales for one-and-one-half percent, but an extra half-percent was a good investment if it improved prices, and the customers thought Father's work worth the difference. The farmer, of course, wanted all he could get, even if it was not enough to clear his debts.

Not all farm sales were forced. Old farmers retired and moved to town or to Florida, estates had to be settled and prosperous farmers had to sell off surplus livestock and outdated machinery.

An auction was a social event, and the seller was a big man, even if only for a day. Father had two gunnysacks filled with tin cups. These were lent gratis to church groups or others who provided lunch at sales. Early on, the lunches were free, provided by the seller and catered by church groups or by the one black, Hays Porter, who was tolerated in our free-state town. I must say for Father that he and Porter were friends on as near an equal basis as they could manage. Porter was not invited to our house for dinner, but Father was never invited to bring his family to the bankers' houses, either. He aspired to the easy ways of assured position, but he never mastered them. Soon after World War I the statement "lunch stand on grounds" replaced "free lunch" on sale bills. Whatever else it did, it freed Porter to serve the sort of food he liked to prepare and which sale patrons doubtless preferred to what a bankrupt farmer could afford.

I believe that Porter plastered the house in which I grew up, and that he was in charge of building the cyclone cellar. I'm confident that he was regarded as the concrete building expert of the community. If he resided with Father and Mother while the house was being plastered, eating meals with the family and sleeping must have tested his and his employers' ideas of propriety. Father, at least, was spared considerable need to be thoughtful by the belief that blacks were biologically inferior. I remember him saying, one time, that their skulls lacked the longitudinal seam which he attested whites had—that their skulls were truly like a species of lump coal we used to buy and which was called *niggerhead* because many of the lumps seemed to have been formed by whorls of vegetable matter and resembled cannon shot in size. Relations between whites and blacks could be pleasant, but they required toleration on the part of the blacks and condescension by many of the whites.

Sale business was seasonal, peaking before the first of March when farm leases traditionally ran out and loans fell due. Sometimes the farmer could hold on until July, when he harvested a last crop of wheat. Everything was complicated by rental arrangements. All farmers planted more land than they owned, and they owed one-fourth or one-third of the wheat crop to the landowner if seed wheat had been furnished by the farmer and one-half if the landowner had provided seed. Many of the landowners were absentee individuals, and some land was owned by insurance companies or other corporations.

In the years following World War I, Father was likely to have a sale on almost every weekday in February and many in March and April besides. Often, he drove twenty miles or more to the sale. Early, a buggy and team were used, with a Galloway steer lap robe for protection. Later, a Model T Ford served unless roads were exceptionally muddy. Tire chains were employed often during wet seasons. When times were good, Father bought a new Ford every year, often selling the old one at a sale. Farmers were happy to permit this. A car on the sale bill helped to draw a crowd. Trading in cars became the custom somewhat later. By the mid-1930s, with dust storms and the Depression, trading in was the order of the day, and list prices were only an anchor for haggling. But by this time, Father had become a town-based automobile dealer and had begun to phase out the auction business.

Throughout World War I, Father was a highly visible public figure. He was too old for the army, and with three small children, was not likely to be drafted in any event. He was active as a gratis auctioneer for almost weekly Red Cross auctions. In these, citizens would donate items such as knit scarves, kitchen utensils or poultry, which would be sold; proceeds would go to the Red Cross. Often, the buyer, who ostentatiously paid an exorbitant price, would re-donate the item, and it would be sold again, and again it would bring an outrageous amount. In this way our farmers, in town for Saturday afternoon, would demonstrate solidarity with the war effort. Such demonstrations of patriotism were especially important for members of the German-Russian community—or, as they later became known, the Volga-German community. Near WaKeeney, most of these Volga Germans were German Lutherans, though in Ellis County to the east, most were Roman Catholics, and in places farther south and east, Mennonites. They were good farmers, and they were prospering, and it was important that they contribute liberally to all war-related causes. Many of them spoke little English, and often were driven to extravagant gestures in their efforts to make the locals understand. Everybody understood very well that one could be ostracized and even have his farm buildings painted yellow if there was any hint that he was pro-German. Though no one spoke about it, credit at the bank and the grocery store were doubtless also conditioned by perceived patriotism during World War I.

In the boom times of the early 1920s, Father was joined in the auction ring by O. H. Olson of Collyer. This arrangement gave the farmer a more aggressive selling team and prolonged Father's career as an auctioneer. Even so, Father recognized a need for a less demanding occupation and was ripe for the venture into the garage and implement business which began in 1925.

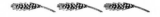

GREEN-BREAKING
WORKSTOCK

In 1916, with America about to become involved in war, horses of all types were in demand. A sound animal four or five years of age and unbroken could be bought for about a hundred dollars. Green broke, the animal would be worth two or three times as much. Father was active in this trade and often had twenty or more extra horses on hand, either green broke and awaiting sale or ready to be harnessed for the first time. John Naiman, a neighbor's son, helped as hired man.

One "experienced" animal and one neophyte would be carefully hitched to the farm wagon. Father would stand in the wagon, holding the reins while John held a long rope attached to the halter of the green horse. When the team cleared the barnyard, Naiman passed the rope to Father and scrambled up the endgate. They would then proceed around a square mile and perhaps to town, where they could negotiate the wagon dump in the Hardman elevator. In the afternoon, the morning's run might be repeated with a different pair if farm work permitted. A day later, the neophyte of yesterday would be the experienced horse and the daily routine started again. Mother probably worried. I stayed at a safe distance. When a wagon team turns sharply, the inside front wheel screeches against a rub iron on the wagon box, and the box rides up dangerously. I was not one to take chances. The number of lessons a horse got varied with his probable worth and his resistance to breaking, but I think the average would be about three hitchings. Horses that showed five or six years age by mouth and which resisted handling were unfit for service. They were more likely to be sold for soap grease and hide than to be returned to pasture.

Mules were in greater demand and were more profitable than horses. At first, Father would have nothing to do with them, but Naiman was a good mule handler and he and Father eventually broke many teams of mules. Thereafter we used mules in the field and found them easier to handle and less greedy at the feed box than horses. Most of them walked faster than our horses and they were less likely to bite or kick unexpectedly. This might have been because they were handled more gently; handler and handled seemed equally eager to keep the peace.

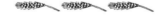

CHILDHOOD

Grandpa Mason died beneath an ash tree near his potato patch on our South Farm—one Grandpa had once owned before moving on to Dodge City and a truck gardening venture, then coming back to live with Father and Mother. Father had bought the South Farm in 1916. My sisters and I, riding our horses, found grandpa when he had died. We had come over from the home farm for a visit. This was the first corpse I had ever seen. Months after the funeral, I remember that Father shared the first of a very few family confidences with me. "Harry," he said, "today I paid the last of Pap's funeral expenses." We scarcely broke stride in setting cedar fenceposts. I preferred to work with Father, though most of the farm work was done by hired men whose pre-Chaucer stories made me laugh, yet feel sinfully guilty. The one type of work that Father and I did together and I did not enjoy was grooming livestock. Father liked to do this on Sundays, and he enjoyed it. I stood holding a halter rope, feeling guilty for working on the Sabbath and trying to be patient.

The hope to move up in the world was clothed, for Mother and Father alike, in making their children's lives easier than theirs had been. Mother did dishes alone after meals. The girls, she said, would have plenty of that to do when they grew up. Father believed in chores. My sister Edith was not assigned outdoor work. My sister Mary was Father's strong helper and boss and driver when we went, behind a slow but patient horse, to school in a buggy. Doing chores with Mary was somehow very grim for both of us, though what we were asked to do was never hard. What was wrong was that we were on our own and responsible. We had to make such judgments as: Have we "stripped" the cow's udder sufficiently? Did a few flakes of chaff and possibly manure from the cow's underbelly foul the milk too much? Without a responsible adult presence, work standards tend to decline, guilt to accumulate and revulsion to set in.

Early in childhood, I showed interest in machines and how they work. Father recognized this and nourished the interest in many ways. I remember one period in the summer, probably in 1916 or 1917, when I would have been eight or nine years old, when Father and I sat, almost every evening, on the concrete wings at the end of our cyclone cellar while daylight lingered, then faded. He told me how a steam engine cylinder turned the expansion of steam into powered rotation of a belt pulley. He described the double-acting piston, the slide valve, the eccentric mechanism, the crankpin and the pulley itself, the boiler, the steam dome, the flyball governor, the firebox and flues and the Penberthy injector which magically forced water into the boiler against its 150-pound pressure. I remember listening with rapt attention, asking questions and sweating through the problem of getting the engine caught on dead center, working the reverse gear—which changed the timing of the slide valve—and in fantasy, sounding the whistle. Father illustrated with pencil drawings, most of them made on pieces of scrap lumber. I would gladly have provided a Big Chief Pencil Tablet if it had been needed. Father even joined me in fantasies of engine operation. He dramatized regulating the throttle, firing the boiler until the safety valve popped, listening for possible irregularities in the exhaust and checking dampers and forced draft. I could tell that he understood steam engines well, and from other conversations that he understood operation of the threshing cylinder, the shaking sieves, fans, return elevators and wind stacker of the thresher.

It must have been in 1919 or 1920 that Father began to take me along when he spent summer Saturday afternoons in town. He would auction miscellaneous items that people had brought to a vacant lot between two stores on Main Street, then spend time in conversation with farmers who had come to town to shop and visit. Probably unconsciously, he was keeping his businessman image shiny. He made it a practice to give me a quarter to spend as I wished when we arrived at our parking place. I think he expected me to find my friends Walter Kline or Harold Burnett and scrabble around in the downtown area. Kline's father owned nearly a solid block of business buildings, including a grocery store, a furniture and embalming emporium, a men's furnishing store, a drugstore and a bakery. On a few occasions I remember that we did roam the back rooms of the stores, but I detected a putting-off in the attitudes of the merchants, and this discouraged me. Merchants did not want to antagonize the son of their landlord, but they probably did not want two or three immature boys barging around in their places, either. Kline was at least a year older than me and aeons more sexually advanced and aggressive. I felt inferior and a bit sinful, listening to his accounts of exploits with girls, even with women mature enough to be thought of as mother or auntie.

I gradually tapered off association with these playmates, and developed a routine of my own. I might spend five or ten cents of my quarter for an ice-cream cone, or I might save the whole quarter to apply to the purchase of a baseball or a fielder's glove. Instead of looking up Kline or Burnett, I would walk up to Main Street, across a vacant lot and an alley, to the back entrance to Irv Woerner's welding shop.

The back of the shop was mostly doorway—a space wide enough to drive tractors or trucks through or to accommodate the damaged projections of machines so that a torch could reach them. As one, two or perhaps three summers passed, the work of the welding shop changed. Repair jobs became less common, and construction projects proliferated. Many of the projects were tanks for hauling gasoline or other petroleum products on highways. The oxyacetylene torches were supplemented with WaKeeney's first electric arc welder. Without taking in its significance, I was seeing the emergence of an early surge in the petrochemical culture and the flowering of one of WaKeeney's few successful manufacturing efforts. Woerner continued to make tanks in WaKeeney until the beginning of World War II; then, I think, he moved to Wichita.

In 1915 or 1916, my two older sisters and I had started driving a buggy to school in WaKeeney. The first horse we drove was named Jeff, a freshly retired mover of Standard Oil Company of Indiana's local tank wagon. It is useless to speculate whether Jeff fell from tank wagon horse to school buggy horse because of old age—he was well past twenty—or because increased demand for gasoline forced his former owner, Charley Sellers, to substitute a team for the one-horse delivery vehicle. By the spring of 1923 we three Mason children had gone through Jeff and two more buggy horses, my sisters had graduated from high school and Sellers had put real horse power out of business hauling gasoline and kerosene. He had mounted Standard's tank on a Reo Speed Wagon. I remember seeing the truck on its side on the road just south of WaKeeney's cemetery one morning when the road was a bit muddy. Apparently autos, mostly Ford Model Ts, had created meandering ruts, and when the Speed Wagon followed these, a partial load of fuel had sloshed from side to side violently enough to lift the truck neatly onto its side, crosswise of the road. I hasten to emphasize that the tank was a company item—not one made by Woerner. Woerner's tanks contained baffles to decrease sloshing, and, I suspect, drivers had learned to drive slowly and carefully with partly loaded tanks of liquid.

I did not invade the social space of the welding shop, but took up station on a stone or cement block near the back door; Woerner, I learned later, was a kind and sensitive man, and he might have had a hand in providing me something that was not a real seat, but usable as one. He and his men must have been amused to see a youngster come, Saturday after Saturday, to the back of the shop, sit on his block of stone or cement and stay there for hours with scarcely any perceptible movement. I did not speak or call attention to myself. Soon Father knew where to find me when it was time to go home to do chores. If he exchanged a wink with Woerner's men, I was never conscious of it.

In 1925 when Father went into the garage business, and for a few years after, Sellers was still delivering gasoline for

Standard Oil; he was our tank wagon man. He had, of course, a truck larger than the original Speed Wagon, and he used a large hose to fill our underground tanks. If the price of gasoline were to increase, he gave us prior notice, and always made sure we were filled up at a lower price. Within a few weeks of the time when he would have had a vested pension from the company upon retiring, he was fired, and a new man took over. By that time Sellers had sustained an injury that left him with a severe limp. Everyone who knew about Seller's treatment by Standard Oil was incensed, but none of us protested vigorously or formally. I like to think that things would be different today—that a certificate of protest would be circulated, that, at least, we would have changed brands of gasoline. But, I don't know.

RELUCTANT HORSEMAN

Somewhere there is a yellowed photograph showing Father's Old Billy, with me, on my third birthday, looking scared on Father's saddle. Morgan must have stepped back out of camera range just long enough to see me achieving a milestone in horsemanship—being in the saddle alone rather than clinging to an older person from behind the cantle or squirming uncomfortably in an adult lap. Vaguely, I remember. There also seems to be an earlier fragment of memory, in which I had fallen to the ground just as horses erupted from the barn door. Old Billy or one of the other horses walked over me, carefully placing its feet so that I escaped injury.

Another early memory involves horseflesh. In the autumn of 1911, Father had moved our family into WaKeeney so that Mary could start first grade. Edith, a year and a half younger, cried at being left behind and was allowed to accompany her; the two of them then went on through grade school and high school together. The winter proved to be one long remembered for its blizzards and snow, and Father's judgment was vindicated. As the house-high snowdrifts receded in the spring, Father began preparations to move back to the farm.

One morning I heard my sisters shouting; there was a colt in the barn behind our rented house, and they rushed out to see it. Hopelessly behind as usual, I churned out of the house on my short legs, but I never got to the barn. I fell on a broken bottle and cut a gash in my right leg just below the knee. I remember Doctor Jones, with his chewed-up cigar, cleaning the wound and sewing it with three stitches. The number of stitches was a measure of the seriousness of any accident. When I whined at the pain, he asked, "Does it hurt? I can't feel it!" So much for clinical humor and the status of children in WaKeeney in 1912. I also remember that when I finally did see the colt, it was at least a yearling. It looked like a slightly undersized horse—the endearing spindly legs had been overtaken by a developing body. Looking back from my eighty-fourth year, I see this as one of a long sequence of indifferent encounters with horseflesh.

Why could my father give up horses reluctantly, Wendell Berry hold to them after nearly all farms were mechanized, and I, from the first, be driven to associate with them only to please my revered father? I feel that the ages-long relationship between horses and men can be roughly appreciated (understood is too strong a word) by thinking servant-master, adversary-friend. The horse's role in the economy is to serve man; the atmosphere that master and horse inhabit can range from grudging respect to warm affection. Berry seems to love his team and they to serve him gladly. Father was the Republican Committeeman, happy with his horses yet anxious that they keep their place. As a child, I could read Father's attitude, but I was both too immature in person and too small physically to build a real master-servant relationship with a horse. I came to horses too young and too small. In summer, I could have had shoes, but I preferred to be barefoot. One of my more painful memories of Dandy, the painted pony, was of his left-front hoof sliding off my bare foot. Flies were abundant around the barnyard, and horses had to brush them off with their tails and muzzles and stamp to dislodge them from thin-skinned front legs. Bare human feet suffered.

Another source of trouble for a young rider was mounting up. Our saddle stock was broke to stand while the rider swung up from the left stirrup but to move off before the rider was fully seated. This kept a short rider anxious. One method we used when away from the barnyard was to find a cutbank—a short cliff of about fourteen or fifteen inches, where the flat plateau of the prairie had slumped from erosion. Prairie land is not as level as it seems to be from an Amtrak car or from an automobile on the interstate highways that parallel the tracks. The railroad was laid out to avoid hills and valleys; highways encroached on the right-of-way, often after lawsuits had been settled. So, in the pasture—the less plowable land of my father's home quarter section—one could find a cutbank, guide the horse to expose his left or "near" side below the grassy crest and mount with an advantage of a foot or so. How much pain and conflict for both horse and rider might have been saved if someone had convinced farmers sometime before 1910 that a rider must be able to mount easily from level ground before being trusted atop a horse.

By the time I was six years old, Father had given up moving to town for the school year, but he had not given up having his children educated in the WaKeeney school; it was presumed to be superior to the slightly closer one-room Chalk

District schoolhouse. At first, Father must have driven my sisters to school, either in his 30 Horsepower Regal or in a buggy.

When I entered first grade, Mary was entrusted with a horse and buggy to make the daily trip. The pattern was to leave the farmyard in plenty of time, loaded with syrup-pail lunches and covered with a warm lap robe in cold weather, and drive to a livery barn near downtown WaKeeney, about a half mile from the school. There, liverymen unhitched our horse and put him in a stall to wait out the school day. We walked, or in later years skated with roller skates on the sidewalks, to the schoolhouse. At first, Old Billy was considered too fractious for us, and Father bought Old Jeff, the retiree from pulling the Standard Oil Company tank wagon. Jeff was somewhere in the far end of his twenties, which is old for a horse, and he was devoted to taking his time. Mary slapped the horse's rump with the lines and beat him with the short buggy whip, but nothing would inspire him to go fast enough to please us. Forcing the pace was the spirit of Manifest Destiny, and though we had never heard the term, we were of a temperament and a degree of immaturity to find moseying a sin.

Within a few years, Father placed Jeff in his second and final retirement and gave us his beloved red bay Billy, the one that had carried him many miles in the saddle. There was a distinct class difference between saddle horses and draft stock, but a saddle horse could double as half a buggy team, or even as a single buggy horse, without losing prestige. Mary's driving record and growing maturity made her a suitable driver for the still fractious Billy. He must have been in his late teens. A Russian thistle blowing across the road would elicit a brisk sideways "shy" but did not endanger the buggy if the driver was alert. Mary always was.

There were two memorable emergencies during Billy's tenure as school horse. The first one occurred during cold weather, when a lap robe was tucked in around feet to keep them warm. In addition, Father heated bricks, which were placed in the bottom of the buggy frame. These were taken out by the livery men and kept on the top of their office stove and replaced for the trip home. I remember that I had been left at home, probably with a cold. Looking out across the prairie shortcut Mary often took on the homeward trip, I saw smoke, then flames rising from the buggy, Billy straining at the bits, and my sisters frantically pitching out burning blankets. Had there not been snow on the ground, the dry winter grass would doubtless have burned. Father was furious at the liverymen but, as I remember it, he wisely decided that it was better for the children's feet to be chilled than to attempt to educate the liverymen as to the proper temperature for foot warmers.

As a bit of history, it is appropriate to say that there was a still more dangerous way to keep one's feet warm in a buggy or in an early automobile. This was the charcoal footwarmer, a sloping-sided metal device that contained, in one end, a controllable vent for draft and, in its interior, a drawer in which two brick-sized chunks of charcoal could be housed. These heaters had such a firm reputation for danger that Father was not tempted to offer one for our use. During one long cold winter we rode horseback to school, employing, in some rotation, Billy, Father's American Saddler Lady Belle, Dandy and a mule named Alice. Alice had by far the easiest gait. She would walk fast cheerfully, but she refused to trot or gallop. The greatest problem with horseback travel during that winter was snowballs, which caught in the hooves of the horses, built up to a height of an inch and a half and had to be removed with a stick (or a schoolbag ruler) to prevent stumbles. On a few occasions I rode behind Mary's saddle, and I can remember having my fingers cramped shut cold, even through mittens, to the extent that my teacher rubbed them to ease them up. The nearest we came to frostbite, however, was when, several times, the middle of a rider's cheek would accumulate a hard spot about the size of a cherry. It usually disappeared in a day or two, but the cheek might be tender for several weeks.

Old Billy's last school caper occurred on the way home on a spring afternoon. We had stopped at our mailbox, a half mile from home, and found a package from Sears or Montgomery Ward, containing a photographic developing outfit, complete with a red chimney for a kerosene lamp to be used in a darkroom not equipped with electricity. We turned left from the mailbox, eagerly unwrapping and inspecting the equipment, and Billy was allowed to trot, as was custom, down Sellers' hill, a short dip in a draw that drained land to the north. Billy saw a tumbling Russian thistle and, probably feeling it his duty, shied to the right. This would not be much of a problem usually nor expose the buggy to danger. But this maneuver was different. The usually docile Billy went into a bucking frenzy, tore up the harness, broke the shafts and turned the buggy on its side. Two of us went out over the left side; Mary went out and under on the right. None of us was hurt, but it took some time to gather up the contents of the package. We were not to know what had caused

Billy's behavior until we got home—a walk of a half mile. Then we could see a spot on Billy's right hind quarter, about the size and shape of a football, where the winter hair had not been shed. Under this spot was an angry blister. When the blister was thrown against the buggy shaft, the pain had been so intense and unexpected that dependable old Billy shed his sanity. The blister's explanation was simple. Our few laying hens ran freely through the stalls of the horse barn, and they had become infected with roup, a throat disease. The remedy for roup was to lace the hens' drinking water with copper sulfate. Their drinking trough had been in the back of Billy's stall. A splash from it had permeated the straw-and-manure bedding of the stall, and the blister had been the result. Regular grooming would have disclosed the blister, but Father's days of devotion to horses had passed, and we children thought of the school horse only as a means of transportation. Wendell Berry's horses would not have found themselves in such a fix.

Old Billy figured in one other well-remembered incident. Up until the Armistice Day parade in 1919, Father led WaKeeney's parades astride Old Billy. On the occasion of Armistice Day, 1919, members of the American Legion had decreed that the parade must be led by a retired officer of the American Expeditionary Force; a local insurance agent, Max Coldiron, drew the job. Coldiron was not at home in the saddle, and there was some tittering when he teetered from side to side on the aging but still fractious blood-bay. He verged on the disgrace of "pulling leather," a term we applied to tenderfeet who were obliged to grab frantically for the saddle horn, pommel or any straps that came handy, in order to stay astride the horse.

There is another story about Doctor Jones, the man who sewed my cut knee. He also had a farm about two miles south of WaKeeney, with Big Creek running diagonally through it from northwest to southeast. It must have been about 1915 or 1916 when he initiated his irrigation project. The idea was to pump water from Big Creek up to an earthen reservoir, then pipe it down to ditches gouged among rows of crops and vegetables. To initiate the project, he seined huge carp from the creek, cooked them and invited everyone in the area for a fish fry. There was a large crowd, the weather was fine, and everyone seemed to have a good time. When it came time to demonstrate the irrigation machinery, I remember that my father was given the honor of cranking up the Heider tractor that was belted to the centrifugal pump. Properly suited and hatted, he brought the crank up smartly, and the tractor burst into song. The pump, however, did not prime, and only a small trickle of water came out. This ended the demonstration almost before it began. Within a few days, the pump was made to work, and pumping to fill the earth-lined reservoir began. It soon became clear that the pond was not going to fill. The geology of the Kansas Plains is complex, and even where there is a saddle of loam and clay soil under the native grass, there may be sand or rock a few feet farther down. Sand seemed to be the undoing of the Jones reservoir. Doctor Jones didn't give up easily. He was pasturing a herd of fine steers for a man named Bob Kirk—a devoted stockman who had worked his way from Scotland to America with a shipload of select Shorthorns. Now, years later, his way with cattle had made him affluent. Imagine the commotion when Kirk found out that his own stock was being used to tramp down the mud in the Jones reservoir.

The round and round circuit of the steers was stopped, of course, but not before proving that the natural bottom of the pond was too porous to hold water. Jones later devised some way to irrigate a few acres, though not large fields. Across the section line road to the east, his neighbor Charles Steinberg, a machinist, had somewhat better fields for irrigation. He succeeded in raising forage crops with irrigation, and those fields were still being irrigated in 1987.

Field Work

Farmers who were oriented toward the town life, that is, who felt more fulfilled in a suit than in overalls, were more likely than others to concentrate on winter wheat as a crop, relegating egg gathering, milking and other barnyard enterprises to women. Some farm women became full field partners of their husbands; they were the ones who had plenty of athletic energy and who felt little threat in being regarded as "coarse," rather than "refined." They probably reasoned that, to be refined, it was enough to shop in town on Saturday afternoons and perhaps attend a dance on Saturday nights. Like their husbands and children, they were captives of the American Dream. That dream, in WaKeeney, was to be realized by planting and harvesting progressively larger acreages of hard red winter wheat.

Field work for this crop began within a few weeks after harvest. The ground was plowed with a share-and-moldboard plow, usually a gang of two fourteen-inch moldboards supported by a three-wheeled frame with a seat for the driver.

Typically, the plow was driven along the edge of the field, counterclockwise, throwing the dirt outward toward the fence, if indeed there was a fence. Once around a portion of the field containing about twelve to fifteen acres, the team was directed so that one horse walked in the furrow, the flat-bottomed ditch that had been exposed by the plow; the plow's arrangement of beams and doubletrees was adjusted to place the plow on unplowed land, neatly joining the inside of the furrow last plowed. Plowing the land then consisted of walking and pulling for the horses, and for the farmer, supervising the depth of penetration of the plow, urging the laggard horses by word (they knew their names) or by slaps of the lines, and seeing that the plow scoured. Scouring meant that the dirt slid smoothly off the moldboard in a continuous stream, rather than sticking to the moldboard and dropping in a rough heap without being properly turned over. Soil that had the proper moisture content scoured easily. Soil that was too wet gave trouble. A plow that had been allowed to rust would not scour dependably until the rust had been worn off. A plow could be protected from rust by a thick coating of axle grease, left on the plow over the winter. This could be washed off before starting to plow, but a few hours' plowing might yet be needed before scouring was assured. Bunches of weeds could clog up the space between moldboards and the beams that pointed them toward their work. Lines of rank grass that had evaded sod breaking along the edges of buffalo wallows could sometimes clog the plow, and they always put a sudden heavy load on the shoulders of the horses. Failing to wash caked sweat from horses' shoulders, or fitting collars poorly, could disable horses.

We usually had six horses, working abreast, pulling a gang plow; shoulders were inspected and washed down with weak saltwater in evenings, and each horse had its own collar. When we lent horses at harvest time to our neighbor, Sam Campbell, each one was accompanied by its own collar, and

Campbell always inquired about each horse's name. Sometimes this was embarrassing, as when a particularly rigid-dispositioned animal was named "Sam," or a mare mule was named "Alice V" for a local piano teacher. It was unthinkable to keep only six animals in the barn or corral in order to field a six-horse team. There had to be two or more spares to substitute for animals that needed rest.

About once in ten days, a working bunch of horses was rotated to a pasture about two miles from the home farm, and a fresh cohort was brought in. All one had to do to get the bunch to pasture was to mount a saddle horse and open the gate in the corner of the sparse home pasture. The horses knew the way. One then had to outrun them and open the lush pasture's gate, then round up the replacements and drive them home. This was usually not hard to do. A few days at pasture seemed to leave the horses and mules bored, and they went back to work quite willingly. This could be said of the field teams, not of the saddle horses. When one of the saddle stock had been put to pasture— a much less frequent event—a half-day of hard riding by two horsemen might be necessary to round him up. Eventually the truant would give up and allow himself or herself to be caught, saddled and bridled with leather taken from one of the chase horses and ridden home. The unsaddled chase horse would then have a turn at pasture.

The practice of laying out a patch to be plowed by plowing around it counterclockwise until the unplowed portion narrowed and shortened as the work progressed had two trouble-some consequences. The soil near the field fence was thrown outward, year after year, until a considerable ridge appeared there, and a corresponding low place was created in the middle of the field. A better plan was to alternate anticlockwise plowing from the outside with clockwise plowing, starting with furrows made along a stretched string set up to define the center line of the field rectangle. In this way, dirt thrown outward one year could be thrown inward the next. One had to do some rough surveying to set up the line, however, and Kansas wheat farmers seemed to feel that they would be retired before the consequences of single-direction plowing caught up with them. With luck, this might be true. There was, however, another immediate disadvantage to the outward-in progress of plowing. Both the horses and their driver could see the job they were doing becoming smaller as the days wore on. The psychologists who wrote my early introductory texts and who denigrated animal intelligence should have spent more time on a farm. The horses not only saw the job nearing completion, they began to walk more and more briskly as they saw the area they were to plow diminish. This led to the one serious accident I experienced in doing farm work with a team.

It must have been July or August of 1924. Father had put a hired man with a six-horse team and a gang (two moldboard) plow and me with a three-horse team and sulky (one moldboard) plow to work preparing the South Farm field for winter wheat. He had inspected the field and decided that we would probably finish the field by mid-afternoon. He had been a little nervous about the job and, before going to cry a farm auction, had given strict instructions to the hired man. He was to stop plowing first, so that my team could finish its land while his team was still hitched to his plow, then finish his last few rounds after mine had been unhitched. He was not to make any substitutions in my team (three slow old mares) to make it possible to get the two teams to finish at the same time. Father was well aware of the urge a team felt to speed up the pace when it saw its job narrowing and shortening. It had not been feasible to have one plow follow the other on a single land, because my team walked more slowly and because my plow was two inches wider than one moldboard of the gang plow. To plow the same land with them would result in unsightly ridges, which would jolt harvesting machinery.

At noon, seeing that my team was getting behind on the job, the hired man committed his first sin. He substituted the mare mule Alice V for my furrow horse. When his team was about to finish his land, he followed through with his second gaffe. He allowed his team to finish, unhitched it from the plow, secured the plow to trail behind the service wagon, hitched the team to the wagon and started for the home farm while I still had about two short rounds to finish. I remember vividly that my team began to trot on the outward trip of the last round and to gallop as they turned to produce the last furrow. Controlling them was impossible. I tried to retard them by dropping the plow deep into the ground, but this only raised the wheels of the sulky crazily above the ground, and I soon abandoned the effort.

My runaways had about two hundred yards to go to catch up with the hired man's outfit. Up over the trailing gang plow they came, coming to a stop with the long lever of the gang plow a few inches from my right ear and a bay mare's right foot nearly cut off by one of the shares of the gang plow. Alice V, the mare mule that had triggered the runaway, was not injured.

Within minutes, Father arrived. He had finished his auction and, apprehensive, had driven out to meet us. Fortunately he was in his Model T Ford. We had no need for more nervous livestock. I remember that Father did not fly into a rage. He did not scold me, and I do not think he said much to the hired man. He did not fire him, I know. I think Father had a settled policy of freezing up and going slowly when matters seemed to be getting out of control.

There was, of course, more to field work than plowing, and plowing itself was seldom dramatic or difficult. Usually the stubble turned over smoothly, the team took the job in stride, and the plowman was free to listen to the meadowlarks, even to daydream. When plowing was done, there was a wait for the fly-free date for our county before seeding winter wheat. This was a date established by agricultural experts to allow time for a stage in the life cycle of the Hessian fly, a wheat pest, to occur under unfavorable conditions and thus give the new wheat a better chance to thrive. Usually, planting was delayed until after school had started in the autumn, so wheat was sown by Father or hired men. Sorghum crops were cut with a corn binder and shocked, and I sometimes helped with putting shocked feed into the mow of the barn or into stacks where cattle could, at first, reach through fences to nibble, and later have bundle ties cut and the stalks thrown down to them from the stacks. When bundles or other hay were to be put in the barn's mow, I, as the "kid," always drew the job of keeping the stuff back from the mow door. This was frustrating, because the two men on a wagon could always throw the stuff up faster than I could move it back, and I felt unworthy and choked with dust to boot. Sometimes Father sowed winter rye between rows of corn or forage sorghum. I drew this job only once. We used a narrow grain drill, which would fit between rows of the row crop. It was pulled by a single horse or mule. The driver had to keep seed in the hopper of the drill and maintain a pattern of turning into unsowed rows. I was too much given to fantasy to do this properly. The pattern of sowing required that the drill turn past at least one unsowed row, in order to give a smooth turn at row ends. It was possible to skip a row without noticing and it was possible to sow the same row twice. That I made both errors showed up when the rye sprouted, about the time school started.

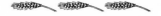

Wheat Harvest
Further Notes

*Thou shalt not muzzle the ox when he treadeth
out the corn.*

—Deuteronomy XXV; 4

When crops were grown to be consumed on the farm,
there was little penalty for having them ripen unevenly over a
period of several weeks. The farmer and his family could take
their own time about getting the crop in. As farms became more
like factories, harvesting, particularly of wheat, was compressed
into shorter times, and the anxiety associated with it was compressed into taller and sharper peaks, until, for the wheat farmer
in 1890 or 1920, harvest on a farm of five hundred to one
thousand acres was accomplished in two periods: cutting and
maturing in shocks or stacks followed, after a break of one or two
weeks, by threshing.

Violent weather—rain, wind or a combination of both—
could wipe out a crop, and as farming on the Kansas plains
gravitated toward a one-crop system, losing a wheat crop could
be disastrous. In our rapidly industrializing society at the turn of

the century, machinery to manage the farmer's risk was invented and marketed, some of the uncertainty of the farm field was overcome through use of the machines, and the focus of farmers' anxiety moved to encounters with their bankers, who provided the credit to buy machines and seed and, by their judgments concerning renewal of notes, determined who "made good" or "went broke" on the land. To understand the changing fortunes of Morgan and Etta Mason, it should be helpful to know in some detail how the separate processes of putting away a crop of wheat have changed during the period of their active work lives—from 1895 through 1920.

Binding, Shocking, Threshing

Cyrus McCormick's 1867 invention, the grain binder, was used on the smaller farms throughout Kansas to cut wheat during the period when sod was broken and open range converted into farms. The machine was pulled through the field by three or four horses, driven by the farmer himself. Its mechanism was powered by a cleated wheel drawn forward with the machine, connected by roller chain to the mechanism of the binder. This mechanism consisted of a sickle, a reel to tip the head-bearing straw back onto a platform and canvas conveyors on the platform and on a short elevator section to carry the cut wheat to the binder head. The binder head itself accumulated bundles of wheat of about eight inches in diameter and tied them with twine. Bundles were accumulated on a bundle carrier, which the driver released to form windrows perpendicular to the sides of the field. The machine typically cut a swath seven feet wide, and, in dry weather, about eight acres could be cut in a day. The farmer, his wife, the teenage children and perhaps a hired man could tip these bundle rows up into shocks, cover each shock with a broken bundle to shed dew or rain and have the crop reasonably secure against bad weather for the two weeks' wait for a threshing machine to visit the community. During this time, the wheat, cut when not quite fully ripe, would mature and dry. Thus threshing might be the only part of the wheat-growing cycle that required transient labor—usually for three to five days.

Heading, Stacking, Threshing

Large farms usually employed the header instead of the binder, and, instead of hiring just one man during cutting, employed crews of eight or ten men throughout cutting. Taking greater risks, these larger farmers allowed the wheat to ripen before cutting, leaving it standing several days more than did those who bound the grain. Then a header, rather than a binder, cut the wheat. Up through the canvas conveyor on their platforms, these machines resembled binders except that they cut twelve feet, rather than seven, at a swath. The conveyors were longer, however, and they fed the cut heads, with shorter straw, into horse-drawn wagons called header barges, which were pulled by two-horse teams and kept abreast of the conveyer of the header. Inside each barge were the driver and a loader. When a barge had been filled, another barge would have been emptied, and it would move under the header's conveyor; the filled barge would be driven to a stack site, where it would be unloaded by its two occupants onto what would eventually be a stack of wheat heads on short straw. As many as four men would spread the wheat out with pitchforks and build the stack—ten feet wide at the base, fifteen feet long, and twelve feet high. Several stacks, parallel to each other and faced so that the prevailing summer wind would blow between them, were placed at a site, with just enough space between them to accommodate the extension feeder of a threshing machine.

To head wheat and get it into stacks, farmers hired transient labor. The cutting-stacking crew included the header driver, four barge occupants and four or more stack builders— nine or ten men in all, who must be fed three meals a day. The wheat, being nearly ripe when placed in the stacks, would mature without undue heating or molding, awaiting the thresher's visit to the farm a week to two after cutting was completed.

When the threshing machine arrived, the farmer's wife, again reinforced by neighbor women, teenaged daughters and perhaps a hired girl or two, would prepare meals for the threshing crew—an engineer, a water hauler, a separator tender, six or eight pitchers who fed the machine from the stacks, four or more wheat haulers and the children who were too young to be employed in the operation directly.

Thus, wheat harvest, circa 1895 to 1900, involved two peaks of concentrated labor and excitement. A lot of critics worry now that farms are becoming factories; the farm has been becoming a factory for a long time.

Combines

Present-day readers are accustomed to thinking of wheat harvest as a single operation, taking place, on a two-hundred-acre farm, in ten days. The machine that does this—the combine—is seldom a prospect for analysis. In 1920, combines were called combined harvester-threshers. The slogan of one manufacturer was, in fact, *Once over and it's all over* to symbolize a new era in wheat harvesting. In 1992, it is common to wonder why wheat harvesting seventy years ago was an exciting time. The excitement has now migrated to caravans of ten to twenty self-propelled, twelve-foot combines, that are trucked from one community to another, northward as the wheat ripens, from the Texas Panhandle through Oklahoma, Kansas, Nebraska and the Dakotas and on into Canada, starting in early June and working until October. The caravans have their own combine-hauling trucks, mobile shops and mobile homes for workers; in a word, their own social life.

Horses

In the wheat binder and its succesors, which dominated the harvesting of hard winter wheat between 1875 and 1920, power to drive the sickle through the stalks, to push the wheat heads back toward it and onto a platform, to carry the heads and their straw supports across toward already-harvested land, as well as to form the grain heads and straw into bundles or elevate them into a header barge—all of this horsepower came from real live horses. The significance of this fact is that the design of the machines and the ways they were used dictated whether or not the horses' work could be called humane—that is, regulated to the horses' needs—or be intolerantly forced upon the patient beasts. All this is likely to be lost on readers whose experience with harvesting wheat goes back no further than the early 1900s. To a concerned reader, it might seem reasonable that consideration for the farmer's beloved teams would take preference over keeping the harvest moving at a brisk clip.

This would be far from accurate. Conditions in the field—roughness of ground, thickness of stand of grain, etc.— vary in such a way that, when the horses most needed consideration, operation of the machine meant that the driver could not allow the animals to slow their pace. When the driver saw a rank patch of wheat he was obliged to force the team's pace so that the machinery produced a vigorous motion and, hopefully, the rig got past the overload rather than getting stalled and skidded through it, leaving bent stalks and uncut grain behind.

In 1880 or 1918, farmers and horses simply had to cope. After that time, gasoline engines could be used to power the machinery. Without urging the team to a faster pace, the driver could speed up the engine and thus the machinery, while the team's pace remained constant. In short, Cyrus McCormick's inventions made demands exactly opposed to horses' needs for compassionate treatment. By 1923, the whole harvesting rig was likely to be gasoline-powered and the horses employed elsewhere. In 1900, a time before most of my readers were born, things were very different, and the difference could account for longer periods of anxiety for farmers, shorter tempers, temptations to enlarge their farms inordinately and a host of other conditions, almost all of them bad. In my experience, immigrants who were bound by church and ethnic customs were more able to cling to responsible ways than their Anglo-Saxon, old American family brethren, who subordinated land stewardship to the ethic of being "up and coming."

SADDLE
AND BUGGY HORSES

In 1944 I worked at Waco Army Air Force Base, research-
ing methods to induce aircrew members to speak loudly and
clearly enough and to hold their microphones closely enough to
their faces so that they could be understood in the din of World
War II bomber aircraft. I was on leave from a job as counselor in
the Testing and Guidance Program of the University of Texas'
School of Education. On weekends I went home to Austin, by bus
or by one of the privately owned taxis that transported thirsty
and lonely airmen from Waco to cities like Austin, where, with
luck, they might be comforted. One of the domestic joys of these
weekends for me was to hold our firstborn, John, on my lap and
tell him stories. By far the favorite story was about the pony my
siblings and I had had when we were young. John promptly
mastered the idea that Dandy had red spots and white spots and
was a much loved and valued creature. From there on, however,
a mythical Dandy developed, little impeded by horsiness. Dandy
could retrieve boxes from the highest shelves in the kitchen
("Wiff his lille paws") and climb the ladder to the tender of
railway engines, escaping burning and abrasion because he wore

"metal mittens." These deviations from ordinary horse-related limitations came to light, of course, when his mother quizzed him as to how a horse could do such things. In what follows I hope to escape the fantasies that bedevil not only children on the verge of becoming two-year-olds, but mature horsemen giving strictly "factual" accounts of their favorite mounts. Where horses are concerned, I feel I'm not love-blinded. For me horses are all right, in their place.

My sisters are four and two years older than I, and my brother is ten years younger; thus three of us reached the age when being more mobile than walking was a priority at about the same time. Our neighbors, the Campbells, provided their children with bicycles, and we envied the Campbell children, but Father did not favor walking sitting down. Neither did he approve of the town banker's solution to expanded child mobility, which was to provide a Shetland pony. Shetlands had a deserved reputation for being contrary, and their sharp, mule-like hooves were capable of causing severe injury, either from kicks or from stepping on their caretaker's feet when being groomed.

Nevertheless, Father was anything but coldhearted. He kept his eyes open for a suitable mount for us, and, one evening, he came home with a small bay-and-white-painted gelding, bought as a three-year-old. Little did he know! I still feel pain in recognizing that Father could be deceived about the age of a horse. He routinely looked at their mouths and pronounced their age with confidence. Dandy, the gelding he brought home, had hooves shaped like those of a typical saddle horse, and this, plus the fact that his mother was a full-sized saddle animal, was enough to convince Father that he had a junior-sized mount for us that was of a proper age to be "broke to ride" and that should have a good disposition instead of the willful attitude of the typical Shetland.

Dandy was put in the horse barn, not with the other saddle stock in a stall, but in the alley in front of the mangers where handlers walked as they doled oats into manger boxes and climbed the ladder to the loft to fork down hay. Within a short time, and with Father's connivance, Dandy soon carried one of us on his back whenever we were not employed currying, braiding his mane into pigtails or combing out and braiding up his tail. All this time, Dandy was growing like a weed. It thus came about that Dandy was never broken to ride in the sense usually implied by Western horsemen. He learned to turn in response to pressure of the reins on the side of his neck, "side reining" and to turn, likewise, from differential pressure on his sides by our legs, though this was a little slow because our legs were not long or strong enough at first.

What he didn't learn was that, when we rode out of the farmyard into the section-line road, we did not always want to go left, toward the mailbox. One may have his own idea of the nature of this bad habit, but my father had a typically Western way of breaking it. He equipped us with a heavy quirt and instructed us to whip him on the side of his face to make him go the way we wished. I think that Father, like many another, refused to acknowledge that a horse has its own idea of the authority of its rider. I do not have any idea how a child can exert authority convincing to a horse, and without it, conflict between horse and rider is inevitable. No doubt each old hand with horses has a different way of teaching a mount to turn whichever way the rider wishes. But at the operational level, all of the methods would depend upon the animal's acceptance of the rider's authority, either through love or fear or a mixture of both. Nevertheless, we enjoyed our pony Dandy, and later, Dandy as full-sized saddle horse. It is only sad that we learned the cowboy's ethic—that cruelty is necessary in one's relation to domestic animals. I still feel badly thinking and writing about it.

Two situations illustrate the trouble that ensued when maturity was not taken into account in giving a rider authority over a horse. The first is a cowboy potato race, put on the Trego County Fair program by a committee that my father chaired. The second is an accident that resulted in a concussion suffered by my brother Howard who was riding behind my saddle. Dandy was the mount in each case.

In the few years before each County Fair engaged rodeo promoters, there were a few cowhand exercises that gained some popularity as grandstand attractions, put on between heats of the harness races. One of these was the potato race. A half dozen men mounted on saddle horses, each man carrying an oak spear made by whittling a point on a binder-canvas slat about five feet long, gathered at a line just behind a strong wooden box about six inches high and perhaps a foot and a half square. This box was empty. Down the track perhaps seventy yards was another box containing potatoes. At a signal, the riders rode to the full box, attempted to spear potatoes, turn and, evading other contestants en route, deposit the potato in the receiving box. Scorekeepers counted, and the race was timed for a duration that I think might have been five or ten minutes. To say the least, it was a demonstration of rough riding.

One year, and I think one year only, a race was designated for riders not more than fourteen years of age. In preparation for it, several boys practiced on Sunday afternoons at our neighbors' farmyard. In these practices, I never scored. The other boys rode plow horses that always seemed to be able to crowd up to the boxes, and Dandy and I couldn't. On the day of the race, I fully expected to be disgraced. I had not counted on the effect of the grandstand crowd upon the horses. My opponents' plow horses were thoroughly frightened and could be kept on the track only with difficulty. Dandy ate it up. He promptly galloped to the full box, put a foot in it and waited until I had a potato speared. Then he dodged the opponents' horses until we had arrived at the receiving box. I think I had to dismount to pull the potato off the stick. Then we went back, speared a piece of a potato and brought it to the proper goal. By that time, the race was over. Dandy and I had put the only potatoes in the end box. We were winners.

I was nearly sick with excitement. Horsemen gathered around Dandy, ready to pay good money for him, in spite of the fact that his front hoof had a line of blood above it, where he had received a wire cut when he found his foot caught in a barbed wire gate. The cut had long been healed, but the foot was chronically tender. The potato race is my last clear memory of Dandy. He had driven cattle—on one occasion Father had arranged for me to ride a few miles west of WaKeeney to help a man named Ossie Long drive a few head of cattle to the WaKeeney stockyards. I remember little about this except that Dandy and I found the place where the drive started and that, when the stock was in the railroad stockyards, Long took me to a restaurant where he ordered the traditional cattleman's breakfast for both of us: steak and eggs, with coffee. If Father had intended it as a rite of passage, it was; I never took horses seriously after about that time.

Howard's accident must have been a year or so before the potato race; Father had sent me to an unfenced pasture north of the farmstead to drive in the cows to be milked, and Howard was allowed to ride behind me on Dandy. Howard was inclined to wild laughter—much more so than to attending to instructions, which were to keep a firm grip on straps holding the fenders of the saddle together and terminating in tapered ends about four inches long. We started, as usual, at a gallop, and, about fifty yards beyond our gate, I felt Howard slipping off, head downward. Foolishly, I grasped at his leg, holding it just long enough for Dandy's rear hoof to catch him on the forehead. I dismounted and gathered him in my arms, and Dandy galloped off for home—a sure signal to a waiting father that there had been trouble. I think I met Father about halfway down our lane. Howard regained consciousness and mobility fairly quickly; I think he was taken to Doctor Herrick, and no serious signs were reported. Howard, now seventy-three, still carries a crease in his forehead. I still remember a confused feeling. I had given proper instructions but had not considered the responsibility to keep my brother safe, no matter what.

A similar accident happened to Mary. She had gone to round up the milk cows in our home pasture, riding the trotting mare, Queen. At a fast trot, Queen gave a comfortable ride. I do not remember whether Mary came to the house on foot or after remounting. What is evident, however, is that, at the age of eighty-seven, her upper lip has a hardened rib in the middle, not very noticeable to the public, where one of Queen's heelless shoes caught her. Queen, apparently over-playful, had fallen suddenly as she shied after a jackrabbit or slipped on wet grass making a turn.

These were not the only incidents—I could go on to two or three more that occurred by the time I was ready for high school. What do these happenings illustrate? Many would say that they show that farming is dangerous.

THE SADDLE HORSE HERD

Though Billy was Father's substitute for Man O' War, two other horses were distinguished more for the prestige they conveyed than for the work they did.

The first of these was Belle Brandon, a light bay with dark accents, which we called Lady Belle. Father bought her from someone who worked in the Kansas City stockyards. She was a five-gaited American saddle horse, and I was impressed when I learned that she was shipped to WaKeeney in her own freight car, probably with a manger, hay and pads to protect her. She had many easy-riding gaits, including a running walk, which could cover ground at more than five miles an hour. Her chief virtue, however, was her ability to stand beside a gate, either a tension barbed wire one or a swinging gate, and cooperate while her rider opened or secured it. She had learned the stockyard lessons well. Nevertheless, she did not fit in, either with the hired men or the other horses. WaKeeney bumpkins would have referred to her contemptuously as *high* schooled. Trained on flat city streets and bridle paths, she was likely to stumble on the mounds of dirt clawed up by badgers around their den holes, or to run afoul of other obstacles. When this happened, her rider was likely to go headlong to the ground. As a redeeming trait, she was steady as an anchor for a lariat. The saddle-horse herd, perhaps five horses, often ran in wild games in the pasture, and a horse of low status was likely to be crowded into the fences. These were made of barbed wire, intended to contain cattle, and severe cuts were the result of contact. Excitable Lady Belle was crowded into the wire twice, getting a deep cut on her shoulder each time. The first cut healed in a short time, but the one on the right or "off" shoulder, received a few years later, was deeper. After months of treatment, ending in a period being supported in a sling rigged by our veterinarian, she grew so emaciated and sad that Father asked the vet to destroy her. I think Father's grief at her passing was genuine, not simply disappointment at a frivolous five-hundred-dollar investment gone wrong.

Dandy, the painted pony, also received a near-disabling wire cut. This was my fault. I had left a barbed-wire gate too far from the adjoining fences while shooing a calf through the opening on foot, and Dandy, tied to the gatepost, got his right forefoot entangled in the gate. His reaction was to plunge, and the result was a cut along the top of the hoof, which put him out of service for a time but eventually healed. The hoof growing down from that area was never smooth, and when he had been ridden too far on hard ground, the hoof line was likely to be smeared with blood. Any wire cut reduced the cash value of a horse dramatically, taking from fifty to a hundred dollars off the best bid at an auction. (One of my sister's welfare clients once referred to an injury incurred by his wife in a way that suggested his devotion: "Ruined her, Mary; scarred her for life.")

Even Billy, in his retirement, demonstrated the connection between prestige in the herd and the probability of injury.

Until his last days he bore no mark of injury except a bold K branded on his near hip, until, long absent from the pasture games, he finally was put into a pasture—and received a cut to a shoulder similar to the one that had done Lady Belle in. He was too old to hold his own in a bunch of horses in the pasture.

Father's second prestige horse was called Queen at home. She was entered in the records of the Trotting Association as *Naida M*. I think Father made the name up. Queen would have been called a *Standard Bred* if she had been properly registered.

Queen was a sorrel, with somewhat irregular white markings, and thus looked a bit too informal for high society. She could trot fast, in spite of all this, and she loved to do it. At the Trego County Fair, she pulled a bike, or bicycle-tired racing sulky and a driver, in such races as the two-thirty trot or pace or the free-for-all. She almost always came in third or better, and sometimes she won. A horse that could be depended upon to place was considered good enough to keep her owner in food and herself in feed, hay and care. Usually, Father allowed one of his driver friends to race her and to keep the greater share of prize money. The driver also paid the entrance fees, which were the main source of the purses.

Unlike many racehorses, Queen, full of competitive spirit for a race, calmed down almost immediately after she was off the track. This made her useful as a buggy horse, though using a racehorse on anything heavier than a high-wheeled training cart was frowned upon by serious horsemen. Following Billy's retirement, Queen became the school buggy horse. Her tenure in this position continued through the school year of 1923, when my sisters graduated from high school. Our road to school doubled, for a mile or a little more, that taken by young Jesse Strain, later known as "Smoky" through his work as a saxophonist in local dance bands. When Mary saw Strain's buggy ahead of us on the road, she would draw up the slack in the reins and cluck to Queen, who didn't need any more encouragement. Queen would elevate herself into racing gear and we would almost fly over the ground. Strain, seeing us coming, would urge his plow horse to 'a gallop, but Queen usually passed him up, reaching the livery barn in a calm though hard-breathing and lathered condition. The liverymen may or may not have told on us, but Queen's tenure as school horse was brief. Mary and Edith graduated from high school in 1923, and when I started high school the next September, I drove a Model T Ford.

Queen's racing days were not yet over. In 1925, an old acquaintance named O. L. Cook, nicknamed Bally, persuaded Father to let him take Queen on the County Fair racing circuit. I think it was at the Hays, Kansas, fair that Queen finished out her racing career; she at least passed out of our ken. Cook, a mediocre driver, had been systematically boxed in by other drivers, so that he could not get Queen into the open to compete with the favored horse—as often as not a black pacer owned by a man named Fred Mangold. The races were run as best three of five heats, and the heats were separated by perhaps a half-hour, while other grandstand attractions were paraded before the ticket holders. As it was told to us later, Bally had been boxed out of competition in the first two heats of a race, whereupon he addressed the offending drivers in a towering rage, saying, "If you try that again, I'm going to shut my eyes and drive right over you." A few breaths later, he fell to the ground, with a stroke. Fred Wolf, a seasoned driver who had not been racing, took Queen to the post in the next three heats, winning them all in a time that I think was two minutes, fourteen seconds—probably the best she had ever posted. Wolf, of course, knew only that Cook had become incapacitated and was carried off to an ambulance. The word that he had died came later in the afternoon. We heard of the incident only secondhand, though from witnesses considered reliable; by that time our family was immersed in the garage business.

THE FARM TRACTOR
COMES TO WHEAT COUNTRY

I must have been past thirteen years of age when Father told me the story of WaKeeney's Gasoline Power Machinery Company (GPMC), which set up a large corrugated iron machine shop in which machinist Charles Steinberg held sway.

Apparently GPMC was an attempt to set up, in the heart of the wheat country, a factory to produce a tractor that would motorize ploughing and seeding wheat fields. Stock must have been sold because the size of the machine shop building and the extensive shafting for machine belts indicated that the lathe, power hacksaw, shaper and mill that Steinberg employed had once been part of a larger plan. According to Father, the effort fell apart when the first tractor model was assembled and tried out. It could, Father said, just about pull itself on level ground, with no load attached. This, I took it, was not Steinberg's fault but that of the engineers who had furnished the design. It may well have been that the promoters of GPM, to use the tractor's name, were more interested in promoting the factory than in making tractors.

In my youth I knew Steinberg as the man who could overhaul auto engines, making the new piston rings from slices of oversized iron pipe with the proper amount cut away and the whole rounded up again. Steinberg also sold gas engines suitable for pumping water when the wind failed, as it once did for a whole month, while a horse-and-a-half gas engine with hit-and-miss governor pumped water for our horses and cattle. Steinberg was an early director of the WaKeeney Band, which employed downtown businessmen as musicians. The Band lived through my youth, often being directed by the high school's Director of Music.

By 1925, I had been recruited into field work with horse and mule teams. From the seat of horse-drawn plows I witnessed the real beginning of the use of tractors as prime movers in farm field work. By 1924, the Wallis, the McCormick-Deering and the Twin City, patterned after the automobile but more rugged, the Rumley Oil-Pull, the Hart-Parr and the Waterloo Boy, resembling scaled-down versions of the gasoline or kerosene-burning engines that had replaced steamers as threshing power, were being seen as our hired man and I drove our six-horse and four-horse teams to and from plowing in wheat fields during late July and August.

In the autumn of 1923 I started high school, and in the fall of 1925 Father joined a pair of WaKeeney bankers in a partnership (he was the visible one, they the silent) to buy an automobile garage in WaKeeney. A year earlier Sam Campbell had bought a garage business from Fritz Staatz and had moved his family into WaKeeney to run it. Campbell was not comfortable in the garage business and wanted to sell out and move to Florida. Father's banker friends and Father himself guessed that Staatz would not be happy until he was back in the garage business.

They schemed to buy the garage from Campbell and sell it back to Staatz, who had broken ground for a hotel across the alley from it. They were only partly right. Staatz wanted a garage to go with his hotel, but when he saw the play developing he conferred with the insurance authorities and found that he could combine hotel and garage in one building. The bankers were stuck with their investment, and so was Father. Worse yet, Father, as front man, was stuck with running a garage. I think he was ambivalent about the garage. Father's principal interest in machines at the time was not in how they worked, but in whether they would make money for him. Still, the garage would provide a channel through which Father could direct my energies. It would also allow him to use the services of Mary, who was his strong helper. Father trusted his own family and was a bit suspicious of almost everyone else.

Father and his backers promptly secured the Chrysler franchise for WaKeeney and the surrounding area. We dealt with the Salina distributor-dealer, Breon-Hudgins Motor Company, which later became Bert Breon Motor Company. Chrysler was just then finishing its first triumphant year.

Two years later, when I had graduated from high school and become a full-time member of the garage venture as a tyro mechanic, a slow-talking but persuasive block salesman for the John Deere Plow Company called on Father and told him that the John Deere agency in town was being closed and invited him to take the franchise. The bait for the deal was the then-new John Deere Model D tractor, presented as the simplest, most fuel-efficient and maneuverable answer to the prayers of the farmer who could no longer produce enough wheat on the land he owned (or could rent on shares) to feed his family and send children of both sexes to college. The salesman had no trouble convincing Father that a man who knew horseflesh could make a killing relieving farmers of their outdated plow teams in trade for the required cash down payment for a Model D tractor, then selling the farmer's trade-in at Saturday afternoon auctions in the vacant lot adjacent to the garage.

I remember my horror at the prospect of mechanical work on something so far from Chryslers and Buicks and Packards; I would later recollect that intuition is a poor guide. In the years up to the Depression's 1929 beginning in the East and its descent on the high plains in advance of the dust storms of 1934, I became as happy with work as I have ever been. Quick help was essential to a farmer whose tractor had died in the furrow; I could usually revive it. When I could, the farmer was truly grateful; if parts had to be ordered from Kansas City, the farmer was usually patient. At eighty-four, I see this period as the one when I was easy in the saddle and on top of the job.

The Garage

Automobiling

Eighty-four years old in 1992, I have seen the replacement of the horse by the tractor in farm fields; the growth, flowering and senescence of the American automobile industry; the decline of railroads as haulers of passengers and freight; and the appearance of all-weather farm-to-market roads in Western Kansas. These roads have replaced, first, meandering ruts, then section-line roads on which farmers worked out their poll taxes and mired their vehicles. Broad outlines of this period have had adequate documentation, particularly in romanticized fiction. Stylized postcards of 1911 to 1916 show wasp-waisted women, well hatted and veiled, riding in the left-hand front seats of early Buicks and Oldsmobiles, protected from dust and grime by cream-colored linen dusters that swept from throat to ankle. My mother wore one of these on the infrequent family outings in our Regal 30 (for thirty horsepower). Father had paid Bud Jones, son of WaKeeney's leading physician, a little more than a thousand dollars for it in July 1911, when I was nearing my third birthday. I can remember reading the embossed REGAL 30 on the hubcaps, which I take to mean that we still had the car in 1914 or 1915, when my progress through two grades of school would have enabled me to read.

Most of the Regal's mileage was racked up taking Father to cry farm sales, or to WaKeeney for groceries and other small items. On summer afternoons we would see our mongrel, Shep, start up the farm lane; one of my sisters would shout, "Poppy's a comin'," and three children and a dog would race out the farm lane and up toward the farm's northwest corner, where Father would stop to let the three small humans into the back seat for a ride into the farmyard. Unless there were Jonathan apples, I would not have been interested in groceries taken to Mother in the house; I would be examining the carbide generator and smelling the slight leaks of acetylene gas that fueled the Regal's headlights. Later, this device was replaced with the safer Prestolite tank, which was turned on and off with a fascinating key. We surviving Masons are fortunate that I was too small or too fearful to find where Father hid the key and turn the gas on. At night, the headlamps created a dramatic spot of bright, safe, yellow light, a wonderland separated from the surrounding fearsome dark.

I remember only one family trip in the Regal. Grandfather Mason and Father's older brother Charley had settled in Dodge City, an even hundred miles south of WaKeeney, where they had a truck garden and small farm, and we went to visit them. The way towns were Ransom, at about twenty-five miles, then Ness City, Jetmore and finally Dodge City. There was an established route that was marked but not maintained as a highway. Three memories of the trip persist. I found that a town could contain hills—a great surprise because WaKeeney was essentially level. Second, the Regal overheated on the way home, and Father stopped near a pond to get water. Part of the Regal's equipment was a folding canvas bucket, with snap-over-center folding braces to bring it from a crumpled disk to about two-gallon size. When Father straightened these braces out, the canvas splintered into shreds. The manufacturer had neglected to impregnate the fabric with anything to resist mold. Father's Stetson did the job, however. It was made to stand up to all kinds of weather. My third enduring memory is of a bull nettle plant, which I saw as an oversized sandbur, and avoided. We doubtless had them in WaKeeney, but I had not noticed them. Here beside the pond where Father watered our metal steed, the plant stood out.

Sometimes Father drove the Regal or rode with one of his WaKeeney friends to Ransom or to Collyer, another nearby town, to attend Sunday afternoon baseball games. Town Commercial Clubs promoted these games and furnished uniforms for the young farmers and clerks who made up the teams. For important games, a team's management would import a professional pitcher from Kansas City, and the opposing team was likely to do the same. Thus the outcomes of the games could present problems for bettors. Would the merits of the regular team members, those of the hirelings, or collusion between the pitchers, decide the game?

Trips to games offered local sports opportunities to test the merits of their cars; it was not acceptable to allow a townsman to pass one on the road. As time went on, John Spena, the local Ford dealer, became the man to beat; thus the Model T revolution came to WaKeeney, where Spena's firm, having graduated from setting up windmills to pump water for livestock, ruled supreme until well after World War I and survived through the Model A, the briefly manufactured B and the early flathead V8. In later years there was to be competition, but before World War I, the Model T was what farmers could afford and what they wanted and bought.

Sometime in 1915 or 1916 the Regal was given a rough coat of paint, the tattered top was removed, and it was driven away to become a used car. Then Father seemed to come home in a different car every day or two, the most-remembered being a Decatur, a noisy machine that I think was air-cooled. I remember it because, when Father put me aboard for the last quarter mile of his trip home, I got a headache. It was my first headache, and I have remembered it and the Decatur together. Soon, however, we had a shiny brass-radiatored Model T Ford, with a box full of buzzing coils projecting from the dash into the front compartment. The steering column and controls were on the left—something other cars were to copy, but the left front panel, where the door should have been, was seamless from dash to the back of the driver's seat. To mount from the left, as a horseman preferred, drivers had to climb over the panel, and they usually did. By 1916 the whip socket, evident on cars earlier than the Regal, had disappeared.

Father, like many who drove their cars hard, tried to buy a new one every year, but this pattern was interrupted when industry was mobilized to implement World War I. I think our last Model T "open" or "touring" car was a 1917. Whatever year it was, it was the one on which nonskid tires were introduced. They were imprinted with the words, one above the other, NON SKID, printed diagonally across the treads. When the tires did not skid, they made an interesting pattern in mud. Along with nonskid tires came a rash of punctures. There were two reasons for this. The NON SKID projections on the new Ford's tires were not molded on top of an already adequate coat of rubber, but were formed by making the intervening and surrounding rubber coatings thinner. A short tack could easily puncture this thin rubber. Second, perhaps as a result of an assembly-line speedup, tacks in the gimp that finished seams in the car's upholstery were not seated solidly, and they tended to be shed by the dozen when the car was driven.

Nonskidding treads got a rough introduction, and the compound-action tire pumps, with one large and one small cylinder, gave drivers a great deal of exercise between curses. Many drivers carried a spare tube, preferably a new one. Father, at least, discovered early that the spare tube should not be carried in its cardboard box. The cardboard would chafe a hole in the tube. That, or faulty cold-patch material, or depleted cement, could make a flat tire a real emergency. Fortunately, on dry roads, one could drive a Model T on the wheel rim, without the tire, though this took a toll on the wheel, to say nothing of the rest of the car and its occupants.

After World War I, new Fords were provided with self-starters. I remember trying out ours repeatedly while Father was away from the barn where the car was stored, until the battery was too low to start it. Father was gentle about reminding me that I shouldn't do this. He explained the whole mechanism of starter, switch, battery and generator. He seemed to know that guilt was sufficient punishment. A battery that was too low to start the car would still furnish current for ignition, and the crank and front choke wire, needed for hand starting, were retained up to 1927, when production was shut down for changeover to produce the Model A.

To save money and to be independent of town garages and their repair bills, farmers learned to do almost everything needed to keep a Model T running. They also ordered, from Montgomery Ward or Sears Roebuck, such accessories as exhaust manifold, front compartment heaters, distributor ignition systems and even sixteen-valve racing-type cylinder heads, as well as special carburetors. Most Fords were plain Janes, but some were real dolls.

Sagging fenders could be braced with baling wire wound over the radiator's refill spout and down to the other fender or the headlight standard. Baling wire is often scoffed at by city dudes, but it is strong, about the right size for many jobs and soft enough to be shaped easily. It truly deserved its wide use. A loom of insulated wires connected the four buzzing ignition coils to a wiping timer at the front end of the Model T engine's camshaft. When this loom became oil-soaked and frayed, farmers replacing it would agonize over getting everything connected correctly. At the top, the loom's wires emerged from the sheath to match the proper coils, but at the timer, at the front of the engine, one either had to know the wires by position and color or to install them counterclockwise on terminals for cylinders one, two, four, then three—often too much of a mental chore for a farmer who distrusted his mechanical ability and was almost mortally afraid of electricity.

One summer at wheat harvest time, our Ford began to misfire, and a transient worker, who claimed experience at the Ford factory in Detroit, offered to install a new timer and loom of wires if Father would provide them. The yokels gathered around fully expecting the braggart to fail. To everyone's surprise, it took the man no more than five minutes to complete the job, with everything secured. There was no misfiring and no need to do anything over. Surely the man must have worked at the factory.

One of the first repairs made to a near-new Model T was to replace the choke wire, which extended through the radiator shell and terminated in a neat ring large enough that a finger of the left hand of the cranker could pull on it while one turned the crank with the right. To start a cold engine, one pulled on the choke until the engine fired, and the cranker invariably bent the choke wire, or "gooser" as it was called, back and forth where it emerged. A "new" choke would be made from baling wire, often not very well aligned and threaded. After self-starters became available, a second choke rod was provided which projected near the driver's right hand. A cold engine still had to be cranked by hand, and the small boy who was to man the topside "gooser" was apt to hear the grunt-punctuated admonition, "Goose the son-of-a-bitch."

Once started, a Model T would run merrily on the low-tension magneto which was part of the engine's flywheel, but cold, stiff oil could prevent the cranker from bringing the flywheel up to speed for an adequate spark. Another condition that could weaken the magneto spark was end play in the engine crankshaft, which would allow the magnet-bearing flywheel to travel too far back, away from the coils that the magnets were to activate. To cope with both of these troubles, the switch in the driver's compartment had three positions: Mag, Off and Bat. A steel-encased set of four National Carbide dry cells could be connected between the bat terminal and engine frame; this would furnish an enhanced spark for starting. Once started, the engine always ran better on magneto. The dry-cell set, called a hotshot, often deteriorated and suddenly went dead on cold mornings. Another handicap to cold-weather starting was too-heavy engine oil. There was a special oil intended for Model T Fords, marketed as Mobiloil E. It was lightweight and contained a bit of lard to encourage the transmission bands that ran in engine oil to engage smoothly. Farmers almost never used Mobiloil E. Long before the owner felt he could afford a ring-and-valve job, oil consumption had led to the use of heavier and heavier oil, until a grade called SAE 50, Special Heavy, was in use. In winter time, it was thick as molasses.

It was standard safety practice, in cranking cold Fords, to place a block in front of one or both rear wheels; otherwise the transmission drag would cause the car to creep forward while cranking and, upon starting, to bore ahead with sinister force. There was a ready remedy for cold weather starting. One could block one rear wheel, jack up the other and place the transmission control lever in high gear. Thus the jacked-up wheel would serve as an additional flywheel, the car would not creep forward, and much of the oil drag would be avoided. The unbalanced rear wheel would cause the car, once started, to bounce merrily on the jack, but I have never heard of anyone being killed by being butted by a Model T. At any rate, after cranking a Model T for a half-hour, one may not value his life highly.

Being run over or butted while cranking a car is distinct as a threat from that of having the engine "kick," which has had more publicity. When a gas engine is timed to run, the spark occurs slightly before top-center of the compression stroke, so that the air-gas mixture will have a small fraction of a second to kindle before exploding to force the piston down. At cranking speed, this may be too much spark advance and cause the engine to start backwards or "kick." When the kick is against a self-starter mechanism, little damage occurs, since the starter is built to stand such strain. A man's forearm and wrist are not designed for this, however, and an engine kick can cause severe injury. There were two mutually exclusive folk remedies for preventing injury when an engine kicked. In one, the cranker was advised to wrap the thumb beside the fingers on the crank, so that a kick would force the handle away from the hand and out of it, thus avoiding breaking a bone by sheer force. The other precaution was to make sure the thumb was "opposed" to the fingers, giving a full grip to the handle, so that the cranker's arm could resist the kick and avoid having the flying handle bruise or break the cranker's wrist. Such logical contradictions in folk lore seem purposeful—to fuel arguments. The best advice, as usual, was more conservative: retard the spark enough that the engine could only start forward. But if the retard was too much, the engine might be unlikely to fire at all. Today, experts have a still better suggestion: keep the car in a heated garage. If one drives in an Alaskan winter, friends tell me, he garages the car without being urged.

HIGH SCHOOL

In the spring of 1923 my sisters, Edith and Mary, graduated from high school, and brother Howard was five years old, too young to start grade school. I, having just graduated from grade school, would enter Trego Community High School in September. The summer found me in an awkward stage of growth. There were the first long trousers and oxfords, not high shoes, for dress occasions. I remember being seized by an insane desire to splash through puddles left by rain, spattering my new trousers and soaking the new shoes. I simply found myself doing this, without plan, and felt embarrassed that Mother did not scold me. Custom and physiology dictated that I should be taken seriously—one was called upon in high school by the title Mister or Miss—and the duties I was given at home reflected this.

Father felt it was time to sell the old Ford and buy a new one, but John Spena, our dealer, did not have a Model T touring car in stock. Father was, or pretended to be, too busy to take the Union Pacific daily local passenger-and-express train, the jitney, to Ellis or Hays, where the needed cars were on hand. He warned the bank of what he was doing and gave me a signed blank check for the amount of the price of the new car. I was to get on the jitney at ten o'clock in the morning in WaKeeney, get off at Ellis, where the train remained for nearly an hour, go to the Ford agency and offer the salesperson the check, provided that the car be equipped with a spare tire. By this time Model Ts came with demountable rims, but the fifth rim was bare. Getting a tire placed on it was something one attempted to do without being charged for the tire and tube; he could probably accomplish this by paying cash without any trade-in. If the Ellis dealer would not accept this deal, I was to go on to Hays, try there, and if there was no positive answer there, I was to return on the jitney, which left Hays after noon and arrived in WaKeeney at about two o'clock in the afternoon.

Today I'm not surprised that the avuncular salesmen in both Ford agencies refused to supply the required spare tire—it was the easiest way to get rid of a youth who seemed so unsure of himself, and who was, besides, small for his age. I returned to WaKeeney by train. Father was crying a small sale on a lot near the east side of town, and I walked there from the depot and reported to him as he passed from one lot of machinery to another. Included in the sale was an Oakland touring car, and Father knocked it down to a surprised Burl Frazier, who knew that Father wanted the car himself and would settle for it. The car was the last item sold, and we prepared to drive it home. Father suggested that I drive. I "knew" how one drove a gearshift car and had driven many miles in a Model T. The first move was to back the car in a wide arc, then drive forward out of the yard and into the street. I did fairly well, only crashing lightly into a hay wagon at the end of the backing up. Father inspected everything for damage, found none, and we proceeded home. He never bought a new Ford after that, though, of course, in the garage business he would own plenty of used ones.

The Oakland became my transportation for the first two years of high school. Youths were allowed to drive in the rural areas at that time; adults did not need licenses. When licenses were introduced within the next few years, persons over fourteen or fifteen were "grandfathered" in as existing drivers at the time of the law's passage. I vaguely remember some sort of driver's license being granted, but I'm sure I was not examined for driving ability. At any rate, there was a short period, before I was licensed to drive in town, when I parked the Oakland at the Trego County Fairgrounds, which were just across the street from the high school. Within a short time I was driving downtown to the post office and the grocery store and offering rides to any susceptible female students I found walking to school alone. Many accepted, but none lingered.

What I was to do when high school was finished was put on hold—I never thought about it, but I'm sure my father did. I was not rugged enough to be a farmer, and Father, at one time or another, had said that the only persons he would be associated with in business were members of his own family, because they were the only ones he could trust. He bought and sold livestock with the bankers, but he knew that they would never consider him a full partner. That judgment was due to change in a short time.

Looking back from 1992, it is easy to see that Father had two sorts of worries concerning me and that he spent time, money and devotion coping with them. Beginning about summer 1922, I began to feel tired most of the time, and as soon as I'd finished with a duty, I tended to lie down to rest. It was then that Doctor Herrick made urine tests and found sugar, suggesting *diabetes mellitus*. He recommended Doctor W. W. Duke of Kansas

City as a specialist in internal medicine, and Father and I made several trips by train to see him, staying at the Rasbach Hotel on Wyandotte Street while tests and treatment were in progress. On the trip made in summer 1923, I became one of the early users of insulin, which at that time was not well standardized, and which was administered by subcutaneous injection about thirty minutes before each of the three meals per day. Doctor Duke's treatment combined diet—a rather thin one—with insulin, so that I experienced frequent periods of dizziness and, at first, double vision. Nevertheless, I began to feel better—if not from the treatment, then from the chance to stay at a hotel, eat hamburgers at a small lunchroom nearby and attend moving pictures with Father in the evenings, at such theaters as the Newman, the Electric and Frank Newman's Royal. Stage bands played between feature pictures. One of them, I remember, was Rudy Vallee's. There was a bit of business in these presentations. In one skit a bandsman donned a fedora, walked across the stage, and, on turning toward backstage, said, "Prosperity is just around the corner." This was, of course, a slogan of the Hoover administration, and the country did not turn the corner. This must have been after 1929, long after our early trips—after, indeed, we had been in the garage business for some time—but the whole period of treatment in Kansas City is seamless in my memory, held together by a sort of unconscious puzzlement concerning my father's long period of patient care for me and the tardy realization that, at the time, I did not really appreciate his sacrifice. I simply took it for granted, worrying mildly about the cost but completely confident that what was needed would be forthcoming.

It must have been in the summer of 1924 that Father bought the Dodge Business Men's Coupe. The Oakland had begun to use a great deal of oil, and Father's automobile maven, the blacksmith Harry Reichard, told him that the car had a hollow crankshaft oiling system, and that these tended to throw so much oil around that slightly worn pistons and rings led to extravagant oil consumption. There was, at the time, little to do about it except to abandon the car. (I was later to learn that 1924 or 1925 was the date for introduction of ventilated oil rings with scraper rings above them, which, after many experiments financed by the automobile-buying public, eventually brought oil consumption under control.)

Following his expert's advice, Father went looking in the used-car places in Kansas City for a suitable automobile. What he bought was a dumpy, hard-riding Dodge that combined some innovations destined to last and some that should have lasted but did not. One that lasted was the all-steel body. The manufacturer had found ways to secure upholstery without tacks driven into wood. An all-steel body could be dunked in an enameling vat, dripped-dry and baked, and the result was a finish that could, when dulled, be restored to brightness by a polish called Common Sense. (It might have been dubbed Hard Labor with more justice.) Another innovation eventually taken up by the industry was a twelve-volt battery. In the Dodge version, the battery was recharged and the car started by a single chain-driven unit, mounted on the left side of the engine. Instead of the crashing sound of a Bendix Drive engaging a flywheel, the Dodge cranked quietly with its chain. If, as sometimes happened, the chain came unpinned while the engine was idling, the immediate increase in idling speed told of the power required to induce the generator part of the unit to keep the battery up. The Business Coupe would go about sixty-five miles per hour under normal conditions, and I never found out how much faster it would go with the starter-generator chain broken. The whole body of the car took on a mellow roar at just over fifty miles an hour, so that fifty became the speed we traveled. Any faster speed was used only for passing.

Perhaps the most important innovation of the early Dodge was use of Budd all-steel wheels. These were discs of strong steel, with lock rings to hold the tires in place, and they worked quite well. Other manufacturers were slow to follow along, however, keeping wheels with short wooden spokes until the late 1920s. All-metal wheels had to wait, generally, until drop-center rims were made popular by the Ford Model A. When tires fit loosely on wheels, as they did in racing machines used on dirt tracks, and on early Model As, it was possible for a strong man to bring the tire beads together opposite the valve stem, push the beads down into the drop-center well and remove the beads, one at a time near the valve stem, without tools. Today, safety dictates that the tire beads fit tightly on the wheels, and, while the wheels have drop-center rims, power tools are used to remove and install tires.

Wheels, rims and tires are among the key elements of the automobile that have undergone progressive change in the interest of more comfortable riding and, incidentally, less shaking apart of the car itself. The clincher rims of the Model T required tires whose beads were formed with strong cotton cords, elastic enough to be stretched over a clincher rim, the profile of which was a thin capital C. The beads were shaped to clinch under the horns of the C; thus, they held the inflated tube in and the tire on the wheel. The 1924 Buicks were, I think, the first production cars to employ four-wheel brakes and the first to use the then-new balloon tires, doubtless named in a takeoff on the aircraft industry, where it had been found that it was cheaper and more satisfactory to cushion landing gears with big fat tires on small wheels than to design elaborate shock absorbers of other kinds—though the steel and spring parts of aircraft suspensions were far from simple.

At any rate, balloon and semi-balloon tires came to be found as advertised features on automobiles. The first way an ordinary mechanic came to notice this was that the sizes of the tires were no longer designated by the tire diameter exposed to the road and the diameter or "fatness" of the tire itself. On the 1923 Dodge, the tires were called 33 by 4 1/2, and a 32 by 4 would fit the same rim. On the first model of Chrysler Six, in 1925, the tires were 20 by 5.77 inch size, and a 20 by 5.25, a lighter tire, would fit the same rim. Wheel diameter, rather than outside tire diameter, designated tire size. Balloon tires shifted the cushioning job toward tires and away from springs and made the tires and wheels, the unsprung parts of the road-to-body connection, heavier. They also required jacks for changing tires to start at shorter lengths and extend farther, to lift a tire clear of the road. Moreover, the heavier tires required reengineered suspensions to prevent both low- and high-speed shimmy, at speeds, for the former, of about thirty miles per hour, and for the latter at something above fifty.

Brave engineers and patient customers and mechanics were still engaged in this problem when our family bought a new

Chevrolet Carryall in 1950. We now drive a 1979 Ford, and it seems to be quite stable, not oversensitive to slight imbalance of wheels and not inclined to shimmy, wander or, with its power steering, be hard to maneuver and park. The newer cars may be even better, but the way from 1924 to 1979 was not easy for anyone involved with automobiles. Tire pressures for models in the 1920s were about sixty-five pounds per square inch; they descended into the lower twenties for the early balloons, came up to about thirty-five for a time, then descended to the upper twenties and lower thirties for the radial tires that are now preferred equipment on passenger cars and trucks. Along the way, inner tubes and the flaps that guarded them from the roughness of wheel rims have disappeared. One can expect, nowadays, to put a set of four new tires on his car, have them balanced and, with good luck and a tire store that will change them about from front to back occasionally, run them until they wear out completely, almost simultaneously. It's a long way back to the NON SKID Firestones of the 1916 Model T, the tacks from its upholstery gimp that could flatten two or three of the tires within a few miles and to the roads that became shining ribbons of meandering ruts whenever there was a good rain.

On My Own as Mechanic

Soon after I graduated from high school, in May 1927, our mechanic Paul Clem was killed in an accident. Clem was our master mechanic. We had a toilet room in which there was a lamp socket screwed to a two-by-four stud. The wiring on the socket was open. When we wanted light, we merely screwed the bulb in. It seems that Clem, depressed and possibly fogged with whiskey, broke the bulb in his hand and fell against the wiring and a cast-iron pipe. Father said that Clem shouted and Father pulled him loose from the wiring. Clem went limp and Father called our physician. None of us knew how to apply artificial respiration. Doctor Herrick came quickly, but not quickly enough. Hall Van Meter, a local telephone man also came; Van Meter tried for some time to get Clem to breathe, but he was gone. I had been at lunch. When I returned, Clem's body was on the shop floor, Van Meter was just rising from it, a few others were standing near, and Father was crying like a child. It was the first time I can remember seeing my father cry.

I was sad for Clem, but I saw no great problem in taking up full responsibility for the shop. Thus when, one forenoon a few weeks later, a call came that a Willys Knight Six had died in the road about six miles west of WaKeeney, I answered the call and towed the car back to the garage. It was a puzzling mechanical situation. There was a fat spark, which occurred at the proper point in the engine's stroke. The valve sleeves, tested for timing through a pipe plug in the exhaust manifold, opened and closed at the proper points. The carburetor bowl was full. I was puzzled but not distressed. I reasoned that one of the fine passages in the Tillotson carburetor must be clogged, so I disassembled it, spreading the parts out on a wiping cloth on the bench. When the whole thing had been cleared with compressed air, I reassembled it and reinstalled it. I made sure that the standard approximate adjustments were made. Nevertheless, I was not much surprised when the engine did not start. My mind had become more and more uncomfortable as I worked. The gasoline smell did not seem authentic. Furthermore, the fluid in the carburetor did not sting my hands. It seemed oily. Finally, truth dawned. The gasoline would not burn in a Willys Knight Six, even in one warmed up and on the road.

It was easy to check out the matter. Gasoline was sucked, about a quart at a time, from the rear tank by a Stewart Warner vacuum tank and dumped into the line to the carburetor. Draining the vacuum tank, filling it with Red Crown ethyl gasoline, and pressing the starter promptly revived the Willys Knight Six. Everyone was happy, though I was a bit confused. Why, I wondered, had I not sniffed out the trouble earlier? The reason, on reflection, was that, up to that time, we had had little trouble with the fuel quality of gasoline.

The owner of the Willys Knight Six had filled his near-empty tank with gasoline at a Derby station on the highway, which now bypassed WaKeeney's downtown area. The Derby

Company was not able to police its stations, and the local man had gone to buying gasoline from the Southeast Kansas oil fields in tank car lots, warehousing it in tanks behind his station and selling some to drivers-by and more to farmers for use in their tractors. The Willys Knight Six became the leader of a sad procession. Within a few days, I was called to start a John Deere Model D tractor that had completed its run the day before but had refused to start in the morning. One started a cold Model D by squirting a small amount of gasoline into each of its two priming cups, then, after rocking the flywheel back and forth a few times, bringing it smartly over top center, whereupon it would start up, blowing merrily from the still-open priming cups. I could thus check the tractor with "good" gasoline as priming fluid and determine whether or not it would continue to run on fuel from its tank. This diagnosis did not take long. When the prime had blown, the engine died.

The farmer did not need to abandon the fuel in the tractor's main tank. The Model D was often fueled with kerosene or diesel fuel. It had a small tank compartment in which volatile gasoline could be kept to be used for cold starting and for warming the engine to the point where it would digest kerosene. What my customer had to do was to rename his fuel kerosene and proceed "by the book."

Back at the Derby station, it is not likely that the station owner had a severe problem. As a tank car of gasoline was sold slowly, the fuel would separate, the heavier fractions settling to the bottom. The first gasoline sold would be relatively light, or volatile—fine for running automobiles or starting Model Ds— but as time went on, the product would become heavier and heavier. Possibly only a fraction of a tank car load would be bad enough to be completely unsalable. By quickly getting in a new supply of fresh gasoline, the whole tank could be stirred up, and who would know? Life might resume without significant interruption!

In the matter of usable gasoline, this story features Standard Oil Company of Indiana as white knight and the local Derby dealer as rascal. This was not the only or even the typical pattern. The proliferation of competing oil companies that sold at retail tended to keep gasoline supplies up to a usable standard. Oil companies installed their own filling stations across the country, and touring motorists, ever suspicious, favored these company-owned stations. The majority of retail outlets for gasoline and motor oil were still, in our part of the country, locally owned and operated in the early 1930s. At the pump, motor oil sold for twenty-five to forty cents per quart, and the ring-and-piston seals of most engines required that oil be added every two hundred miles—more often if the engine was worn.

At our garage, Polarine, Standard Oil Company's brand, sold for twenty-five cents per quart; Mobiloil, advertised and marketed by Standard of New York, brought thirty cents. Quaker State, refined from Pennsylvania crude, went for thirty-five cents. In carload lots of about thirty-five hundred gallons, Polarine could be bought for fifty-five to sixty-five cents per gallon. Each brand was dispensed from its own company's distinctively marked fifty-five gallon drum by means of a pump that got up a quart at a time. A funnel-hooded quart tankard, held under the pump spout, assured the buyer of full measure. As time went on, it became obvious to customers that the barrels of

premium grade oil were less freshly painted than those of
Polarine and its cousins. That my father cheated by refilling other
brand barrels with Polarine kept me near depression. I could not
quit; there were no other jobs. I could not confront him. Indeed,
he was no more culpable than any similar merchant up or down
the highway.

Beginning with Mobiloil and Quaker State, oil companies began marketing motor oil in quart cans made of treated
cardboard, with crimped tin ends. Guilt, all around, probably
prolonged use of these containers, so that it is only within the
past few years that more convenient and rationally shaped
plastic quart containers have come into use.

Other aspects of automobile maintenance could be sub-
jects for essays, almost all of them sad. Factory-sponsored
dealerships are supposed to guarantee good workmanship at fair
prices—sometimes the guarantee holds, sometimes it doesn't.
Automobiles evolve into more and more throwaway units, including throwing away the entire machine before the owner is
really acquainted with his new car. More and more, we see highly
advertised new products—seldom do we find contented owners
or dealers. Quality of life seems to evade us.

THE YEARS WITH DEERE

In the first few garage years, we did a brisk business. Father had many acquaintances; we had many of the old Staatz and Campbell customers and Highway 40 brought summer tourists past our door. Mary came aboard as telephone answerer, bookkeeper and, more and more, as collector of overdue bills. I worked after school and during the summers. It was a surprise, then, to have Father take on the John Deere agency in 1928. WaKeeney had a Farmers Union Cooperative that had sold Deere machinery, but it was were phasing out that part of its business. Deere had recently bought the Waterloo Boy tractor company, of Moline, Illinois, and in its factory brought out the John Deere Model D tractor. The Model D featured a horizontal two-cylinder engine, fully enclosed driving parts and a degree of simplicity of operation and repair that was remarkable. There was no part that could not be removed, replaced and adjusted in the middle of a plowed field with tools one man could bring to the job. Never were more than two men required on the job, and I finally learned to manage even the heaviest jobs by myself if help was not nearby. In spite of this, I was deeply disappointed, at first, in becoming a tractor mechanic and one associated with an un-even-firing tractor at that.

My initial discouragement did not last long. Father sent me to deliver new tractors and to answer service calls. Whenever there had been trouble I knew that it was my duty to stay with the farmer until the machine was doing its job. Thus, I had many hours in the country, which I soon preferred to the shop. A tourist might fret and twitch when he saw an immature youth attack the ignition or carburetor on his Packard, but a farmer whose equipment was stalled rejoiced when he saw me coming and watched me work, not to criticize but to anticipate what I needed to speed the repair. When the repair held and the tractor had gone around the field a time or two, he was truly grateful.

Deere's salesmen, called block men, showed good judgment about what machines to stock and did not try to overload us. This was important because machines carried over past the season of use were covered by notes, signed by Father, that would begin drawing eight percent interest at a certain time. Farmers often got bargains when dealers were faced with the prospect of paying interest through the winter. Two John Deere mechanics also answered field calls; they were experts and good men. One was Sam Sechler, who answered when a six-hole corn sheller we had sold began to give trouble. He timed the sheller as it filled a fifty-bushel wagon, and when this took about seven minutes he was unhappy. Inspection showed that the custom workload had worn out the shelling parts; the owner had been so busy making money that he had lost track of the amount shelled. Furthermore, his laborers were feeding the machine with wheat scoops instead of wire-bottomed ones. Our part of Kansas did not normally produce corn crops, and farmers did not have the sieve-type shovels that would have let dirt fall to the ground instead of going through the machine. Sechler was an affable man, and without trying he made friends for the company. Another mechanic who helped us a great deal was named Duncan. Not as breezy as Sechler, he still had a keen sense of humor. On one occasion he came out from Kansas City to help a German-Russian customer's two grown sons assemble a big combine they had bought. It was a sixteen-foot machine, with an extension to the cutting platform that enabled it to cut a twenty-foot swath. As the three men worked, Duncan observed that the two boys used their Low-German dialect to converse with each other in his presence and made him feel invisible. After the boys had made a few remarks questioning Duncan's integrity and perhaps his ancestry also, Duncan asked one of them for a match—in the dialect. That ended the German speaking, he told us later.

Father hired many laborers from time to time to set up plows and other machines, which came by freight, partly assembled. One of these was Bill Markle, who did a good job at assembling, but had wild ways with our old cars and trucks. We kept some farming operations going during these days, and when Bill went to haul some hay at the farm, I tensed up, waiting for the inevitable call. Something, from a rear axle to a doorpost in the hay barn, was sure to give way. Markle also entertained us with stories of his cowboy-like deeds. To my surprise, he matured, moved to Denver and got a job with the Deere distributor there. He got married, and when my brother Howard visited him at work in the 1950s, he was in charge of crating machines for shipment to dealers.

It was Mary, however, who anchored our relations with farm customers and the plow company. Many customers were

German-Russian immigrants. They had trouble with our language, but Mary could take the two-inch thick Plow Works Catalog, go over the illustrations with them and find the parts they needed. Rail service was good then, and a quick letter to Kansas City could bring the part back on the local jitney train within a few days. If there was a real rush, Mary used the telephone and, with an early call and good luck, got the part the next day.

When all else failed, Mary and the farmer descended upon me and asked if I could weld the broken part. I remember one evening when I gave up my dreary mooning over some indifferent high-school beauty and welded each of about forty small teeth to a cast-iron corn planter gear. The machine was hopelessly out of date, and only heroic measures could save it. I still remember my growing wonder as tooth after tooth flowed into a smooth bead under the torch. Welding cast with flux and a mild steel rod was not unheard of, but I knew that I was a sloppy welder, and that I was working above my skill level. Nevertheless the gear worked and with very little hand filing at that. For a few minutes I felt like Wally Warren, Father's legendary threshing engineer, who could always pick up enough junk at the scene of an emergency to fashion a fix.

A Fire

When Staatz built the garage Father and his partners later bought from Campbell, he had equipped it with a steam heating system, intended to burn stoker coal. There was a high chimney that looked sturdy. The steam heat was too expensive to maintain and was not needed in our time because automobiles were no longer painted with varnish, but were sprayed with lacquer, which could stand the weather so that few were stored in the garage. We had a stove for the office area that was connected to this flue. There was also another stove, with its own flue, for the shop. In order to give twenty-four-hour service, a family member always slept at the garage. This was usually Father. When the weather was cold, he often started the fire in his stove with a section cut from a worn-out tire. This would burn rapidly and warm up the office on cold mornings.

The evening of November 28, 1929, was exceptionally cold, and most of our customers had not yet fortified their cars' radiators with alcohol. Ethylene glycol was not yet popular as an antifreeze. Consequently, our storage space was jammed with cars. Father rose early and started the fire in the stove in his usual way. The red-hot stovepipe did not worry him; it was joined to a flue presumably adequate for a large boiler. As we learned later, this flue had been built from one course of brick and mortar, without flue lining. I saw the cloud of smoke soon after I left home, which was about six blocks away. More and more I realized that our garage was the victim. By the time I reached the scene, firemen were packing up to leave, the sidewalk was littered with broken plate glass, and Father was standing alone. I saw Father cry for only the second time.

He told me years later that the next day he went to his bankers, put the keys to the now nonexistent garage on their desk and proposed that they all take their loss and quit. The insurance money would cover much of the investment, and in 1929 it was better to have money than real estate. The bankers would not listen, but insisted that he plan immediately to rebuild "bigger and better than ever" in the spring. With all their presumed contacts with Eastern finance, it is hard to understand why they did not know better.

Perhaps it is not so surprising. Now, people I know in Kansas seem much less pessimistic than those who speak from either coast. Builder Solomon Deines made us a fine garage, about 60 by 120 feet, with much plate glass, and a clear span across the building, provided by a bowstring trussed roof. I think it cost less than ten thousand dollars, and Deines took a new Chrysler as part payment.

At the time of the fire, there were four new Model D John Deere tractors stored near the east side inside the garage. As the place burned, volunteers attached ropes to three of them and pulled them, one at a time, out the back door. The fourth, pursued closely by the fire, became wedged inside the doorway and had to be abandoned to the flames.

Father threw up a small sheet metal building on a corner of the property, far enough from the garage site that it would not interfere with Deines's men, and equipped it with a few power tools, an acetylene welding torch, an air regulator and a DeVilbiss paint gun. We unearthed our air compressor and found that its controls, tank and compressor unit were sound. The burned-out motor was replaced temporarily by a one-cylinder John Deere engine of one and one-half horsepower. Father and a hired man quickly drove three of the "saved" John Deere tractors to our farm and set me to repainting them with official green John Deere paint for the tractor proper and yellow for the wheels. The hoods that covered the fuel tanks were removed and repainted, and the Deere trademark, a deer bounding over a log, was retraced in yellow with a small brush over the logo, which could still be seen under the green recoating. I thoroughly enjoyed the work. To farmers looking for a bargain in a tractor saved from the fire, Father made up a story and stuck to it—the tractors had all been returned to the factory in Moline, Illinois. Somehow, I did not feel guilty in supporting this lie, as I had in being involved in Father's substituting of Polarine engine oil for Mobil oil. I was not analytic about it. I enjoyed painting, and I gloried in remaking the Deere trademark on the hoods and redrawing the raised cast-iron letters on the upper tanks of the tractor radiators. Later, I marveled that farmers never seemed to notice the difference between my indifferent decorations and the ones they covered. It never occurred to me that a potential customer might realize the truth but not mention it, being reluctant to upset the delicate balance between a generous trade-in allowance on an old tractor and a too-close inspection of the "new" Model D. I had not yet learned that a mature person does not always blurt out everything he knows.

With three tractors restored and ready for sale, the question of what to do with the fourth followed. The distributor block on the magneto was scorched and cracked. The fuel tank had ruptured. The radiator, which consisted of cast-iron upper and lower tanks and a gasketed and bolted-in core of tinned pipes, looked hopeless. A minor problem lay with the steering mechanism. A cast-iron segment, which engaged a worm gear on the steering shaft, had been broken when the attempt to pull the tractor from the fire had failed. Father decided to replace the magneto and fuel tank, have the radiator core remanufactured by the Clark Radiator shop at Hays, Kansas, replace the steering parts, repaint the machine and sell it as new. I took on the job without a qualm. It was harder than the earlier restorations, but not impossible. By this time I was ready to draw the deer jumping the log, almost without guidance. (There was a heavy paper stencil that could have been used, but it involved lettering that was obviously stenciled. It did not seem good to make this one machine look so different from the others.)

Father was blessed with luck. His customer for the fourth tractor was a man nearing old age, probably a grandfather whose children had flown before he was quite ready to retire completely. He bought the machine and loved it. I had slipped on only one important point. The grease retainers on the front wheels had been scorched, and I had not replaced them. Dirt worked into the wheel bearings and ruined the inner bearing cones and cups; these and, of course, the grease retainers, had to be replaced at the end of the first season. I do not know for sure, but I think the customer was charged full price for the repair. There, at last, I felt a measure of guilt. I had made a mistake for which I was not free to take the blame.

The manufacturer was not immune to blame. We had trouble, in later years, with front-wheel grease seals and bearings on other Model Ds. This was due not to poor design, but to the addition of grease gun nipples to the hubs of the wheels. Conscientious owners would service these fittings, often with power-operated grease guns, persisting until grease oozed from the insides of the wheels, stretching the grease retainers and making a path for dust to penetrate to the bearing roller assemblies. In another rash of troubles, belt pulleys, which formed part of the tractor's clutch mechanism, would become unstable and wobbly. Often, the first indication was a broken clutch driving plate, a fairly expensive part. Investigation invariably showed that a large bronze bushing, pressed inside the drive gear that had been heat shrunk onto the pulley hub, had come loose and migrated about three inches from its proper place. I repaired these by drilling small holes through the pulley hub, counter-sinking short machine screws into them and threading the screws into the thick bronze bearing to hold it in place.

After a year or two, the tractors began coming from the factory with the bushing again firmly in place. The trouble was past. Much later, and informally, I learned what had happened. In early models, the bronze bushing had been pressed into place inside the pulley bore before the gear was shrunk over the outside. A change in order of assembly had been made, so that the outside gear was shrunk on first and the bronze pressed into place later. Shrinking the gear onto the hub had left a slight ridge in the pulley bore. This ridge had scraped the bronze, shavings fell inside a space where they were not obvious, and the bronze fit loosely as a result. Had I been more literate, I might have written a memo to the factory. As it was, I merely devised a way to protect our customers and kept silent.

Iso-Vis

While farmers and merchants were going broke in the Depression, the consumer products departments of major corporations were busy inventing ways to stimulate business. One of these efforts resulted in Standard of Indiana's new motor oil, Iso-Vis. According to the salesman, the product was superior because it was pre-diluted. It contained low-grade distillate, similar to diesel fuel, homogenized with a lubricating oil similar to—but, of course, superior to—Polarine. As a result, cold weather did not cause it to thicken, and since it had already been diluted to saturation, further dilution due to over-rich carburetor adjustment or over-choking to start cold engines would simply gas off, escaping through the engine breather. The salesman, who displayed his thirty-second degree Masonic ring prominently to cement solidarity with Father, sold hard, and we, along with other garages and filling stations, kept Iso-Vis for sale, prominently displayed.

Luck was not kind to the Iso-Vis program. A cold, wet and slushy winter soon resulted in accumulations of water vapor in crankcases; the ventilation systems then in use were not adequate, and Iso-Vis turned into crankcase sludge. Iso-Vis was supposed to allow longer intervals between oil changes, but engine damage, either from seizing of parts in newer engines or rapid oil consumption and loosened bearings in old ones, occurred before car owners had time to consult the stickers on dashes or in door frames that warned them it was time to change the oil.

One by one, Standard outlets up and down the highway rolled their Iso-Vis pumps back from the curbs and complained— often by letter as well as through the salesman—to each other and the company. Moreover, they refused to pay invoices for the product. I remember that Kip Disney of Ellis, a close relative of Walt Disney, spearheaded the complaints.

Iso-Vis disappeared from the market, but its ghosts did not. The old designations for body, or thickness, of motor oil (light, medium, heavy, special heavy and extra-heavy) were replaced by specifications laid down by the Society of Automotive Engineers, and the numbers SAE 20, 30, 40, 50 and 60 became the designations used in advertising and trade. Improved crankcase ventilation has been incorporated in the engines of newer cars, and more sophisticated ways of keeping oil from thickening in cold weather have been worked out. Now it is possible to buy motor oil designated 10-40, or 10-30, to signify that in cold weather the oil imposes little or no greater load on starters than SAE 10, while in hot weather it "stands up" to provide adequate lubrication.

Along the way, pistons of light alloy that neither collapse readily from high cylinder temperatures nor wear out prematurely were developed, and, more significantly, combinations of compression ring, scraper ring and ventilated oil control ring were brought into use to seal engine cylinders and reduce oil consumption to near zero. For a few years, it was possible to extend the life of an engine considerably by replacing piston rings with new ones, designed for higher cylinder wall pressure and featuring expanders between the rings and the bottoms of piston grooves. These operations are still physically possible, but unless the automobile owner has the proper skill and can either tolerate grease or be delicate enough to work clean, the operation is not affordable. When one's car begins to guzzle oil, one looks for a less-used car to replace it or bets his job against the finance company and buys new—Japanese, Swedish, German, American-Mexican or American-Canadian, perhaps even presumed-to-be one hundred percent American.

WILLYS AND CHRYSLER

Part of the effort to "keep America strong" after World War I was to design and manufacture the combat vehicle known ever since as the Jeep. Not being privy to the councils that set up the Jeep project, I can still make a fair guess at what must have gone on. When shooting stopped, there was an inevitable dislocation. Workers and factories faced layoffs and corporate managers and stockholders faced shrinking business. Rescue had to be mounted, and the name of it was *Keep America Strong*. Defense procurement authorities saw opportunity, the Willys Overland Company saw salvation and citizens of cities in our heartland saw a soft landing from the World War I boom. The Jeep needs no defenders, and it, itself, plays only a minor role in the story I wish to tell. What I am concerned with is the way in which the Willys Overland Company, makers of Willys Knight automobiles, found in the Jeep a way to convert from war to peace.

For as far back in my youth as I can remember, newspapers had carried advertisements for the Willys Knight automobile, extolling its patented double-sleeve valve engine and predicting that, at a magic year, when patents ran out, all good cars would have sleeve-valve engines. The smart buyer could have a sleeve-valve engine that "improves with use" right now, however, simply by buying from his friendly Willys dealer.

The magic year came and went, and no stampede of manufacturers to the sleeve-valve idea occurred. Any mechanic who had had the oil pan off a Knight engine and had inspected the brute in earnest could have told why. Instead of a camshaft, the Knight had an extra, small crankshaft, with two tiny connecting rods for each cylinder, wrist-pinned at their upper ends to cylinder sleeves that ran at one-half main crankshaft speed and opened and closed intake ports on one side of the engine and exhaust ports on the other. Sleeves were constructed and timed so that the four strokes of the conventional engine—intake, compression, power and exhaust—would take place in proper order, once every two revolutions of the main crankshaft. Granting that it might work efficiently, such an engine obviously would be much more difficult to manufacture than the ordinary poppet-valve engine, and repair, which would eventually be necessary, would be difficult.

Though Willys Overland engineers were much in evidence in the design of the Jeep, it did not have a sleeve-valve engine, but a compact, responsive, and not overly heavy engine of conservative conventional design. This engine, and a small automobile designed to use it and to sell for very little more than a Model T Ford, were designated the Whippet and placed on the market. Without the promotion and financing available to Ford, the car sold sluggishly, though the few brave buyers I ran into in my early days in the garage were well satisfied, and I was impressed with the clean design and good performance of the cars.

All this must have occurred before and perhaps up to 1925. Meanwhile, Willys Overland, bereft of any sales surge inspired by the advantages of their Willys Knight cars, continued to produce them. The poppet-valve, Jeep-derived Whippet kept the company in business—just barely. Then in 1927, potential rescue appeared. Ford shut down its factories to retool for the Model A, and the market for a replacement car was open. With better luck, Willys might have, in sports parlance, gone ahead and stayed ahead. It made a brave try, and, at the behest of Ward Marshall of Salina, the Willys distributor, and with the tolerance of Bert Breon, our Chrysler distributor, we of Mason Garage took on the Whippet-Willys-Knight franchise.

What we had to sell represented a major merchandising effort on the part of the manufacturers. There was, of course, the low-priced Whippet. Then there was the Willys Six, a poppet-valve-engined car to fill the thousand-dollars-delivered slot. Finally, we had a newly designed Willys Knight Six, made to offer modest luxury and performance-related sex appeal to those who liked to spend a bit more. We received Whippets in freight-car loads, six to a car, and sold them readily. Willys Six and Willys Knight Six were driven out from Salina and placed on display. They, like any higher-priced car, sold much more slowly. All cars were priced to the dealer without bumpers so that these necessary appendages could serve as bargaining chips between buyer and dealer, as could the spare tire, which was also an extra-price option. Every buyer, of course, got bumpers and a spare tire, but a canny buyer without a trade-in and with cash could get them free. Poorer or less wily buyers paid for bumpers, spare tire or both. Financing was done, I think, by the Commercial Credit Corporation, and once the proper references were given and the proper papers were signed, the dealer went gleefully to the bank to pay wholesale cost plus freight and to put what he regarded as profit into the account that never seemed to contain enough to cover the overhead.

Much of this overhead might consist of service to the new cars. Transmission grease, engine oil, a few gallons of gasoline, initial adjustments of brakes and a three-hundred-mile tune-up were included in the service, and we, as dealers, exerted ourselves to keep customers happy—and not inclined to regard their purchases as lemons. This was possible in most cases.

The production cars did embody good design, but the production for 1927 and following years showed poor quality control in manufacturing. The timing chains on later Whippets, which on the older models gave little trouble, would promptly become loose, ticking and tinkling on the chain covers and needing not just one adjustment, but repeated ones. Some chains wore through the thin covers, so that chain, sprockets and cover all had to be replaced early on. The power handling engine parts usually stood up well, but when one had to remove the cylinder head for any reason, he was likely to find lathe marks on cylinder walls and, in a few cases, in both the Whippet and the Willys Six, two sizes of pistons, rings and cylinder bores in the same virgin engine. Moreover the brakes, activated by rods open to the Kansas mud, tended to work poorly and be hard to equalize between sides of the car, and Whippet brake drums were often so much off center and out-of-round that a typical stop could be a distinct wriggle, followed by a lopsided squat.

In redesigning the Knight engine with six instead of four cylinders, the engineers encountered troubles that were not eliminated in pre-production tests. At the garage, we were accustomed to find vibration in engines to be caused by poorly balanced parts, installed at overhauls. Here, in brand-new engines, there was inevitably a very noticeable up-and-down vibration that occurred when the car passed from about twenty miles per hour to higher speeds in high gear. The cause was not unbalanced parts, but unfortunate harmonics resulting from the change from four to six cylinders. Ordinarily, a six is more stable than a four, but, in these sixes, there were so many uppings and downings of reciprocating parts that nothing short of a very thorough analysis could have foretold the result. For conservative buyers, a demonstrating salesman could usually accelerate the car past the critical engine period before shifting into high, and the buyer, often as not, found nothing amiss. Nevertheless, the bobbing up and down tended to prevent the development of the sort of man-and-machine relationship that could build owner enthusiasm. Furthermore, Silichrome steel exhaust valves, and later, hardened valve seats and, still later, hydraulic valve lifters, removed the faults in poppet valve engines that had made the Knight a possible alternative in the first place. By the time the "patents had run out" on the Knight engine, protecting it from patent infringement had become moot.

We did not have to fight the battle of the Whippet revival for long. In 1928 the Model A Ford hit the roads, and only a few uncomfortable souls could be content with a Whippet.

We did not sell Willys Overland products after the new garage was built following our fire. Chrysler brought out a Six, priced at about one thousand dollars retail-delivered, but we sold few of them. One we took for a family car carried Mary and me for a vacation with the Methodist preacher's large family to the Colorado Rockies near Longs Peak and Estes Park. We were gone about two weeks, traveling most days. En route I repaired the preacher's Dodge Victory Six, which had been recently over-hauled at a competitor's good shop and, which, surprisingly, tied up the shaft that held its ignition unit and also drove the oil pump. Probably the engine bearings leaked less oil, and the vertical shaft did not get its usual lubrication. At any rate, we put the car up on a couple of rocks, took off the pan, got the seized bearing loosened up and welded the twisted-off drive finger for the oil pump. By extreme good fortune, the filling station in Burlington, Colorado, behind which we blocked up the Victory Six, had a proper distributor cap to replace the one that had been shattered when the ignition unit had turned suddenly as its shaft seized, and we were on the road in a few hours. I was a hero and tried to be modest about it. As the reward for my work, the musically sensitive preacher's young remained quiet throughout the whole trip, while I rent the air with my tenor banjo and mournful singing of popular songs around campfires. We almost reached WaKeeney without incident on the return trip. About three miles south of town, the Chrysler Six's spare tire gave out and we had to telephone the garage to send out rescue—or perhaps the preacher's car went on in and alerted the mechanics at the garage. As I remember it, the Chrysler Six never really recovered. It had started using oil, the brakes were shot, and, at about twenty thousand miles, its days of glory were over.

Chrysler did make a good car in the two-thousand-dollar-delivered class in 1930 or 1931. We sold one to Claude Hardman, WaKeeney's leading lumber dealer and grain elevator owner. It was a straight eight, and it featured one of the early all-steel bodies used on first-rate American cars. Our telephone lineman friend told us of one he had seen that had rolled several times on the way off a road fill, was righted and driven away, not very much beaten up. The car we sold Hardman had one glitch, however, caused by faulty design, and cheerfully corrected by the factory—provided the dealer installed the necessary parts. The original design had aligned the drive shaft with the rear axle unit so that the rear universal joint was almost in a bind. The drive pinion slanted upward too much. After we had replaced the universal joint a time or two, a small package, containing two shingle-shaped steel shims, arrived from Chrysler. All we had to do was to block the car up, unbolt the long U-bolts that held rear springs to axle, insert the shims, which corrected the pinion shaft alignment with the car loaded, and reassemble everything. With air-driven wrenches used in present-day shops, it would have been a cinch, but in WaKeeney, it was a half-day's work. The operation was completely successful, however.

Doctor Coil

It must have been in July 1927, that Cale Bingham died. He had been an uncomfortable youngster throughout my grade-school years, and in high school puberty only increased his problems. I, myself, was not well adapted to sports and thus somewhat cut off from male activities, but I had managed to go my own way, avoiding contact sports, trying all the distances short of a mile in track and being tolerated at practice and in the locker room. Bingham had not been so fortunate. On the senior sneak day in spring 1927, he was taken along by a group of boys to swim in a deep pond on Big Creek, in a pasture. A Buick sedan had been backed up to the pond bank, and the boys had climbed to its roof and dived into the water. When Bingham's turn came, he must have been near paralysis with fright, because he merely collapsed headfirst into the water, striking much too close to the bank, with a force that broke his neck. May passed into June, and he lay at home, conscious only part of the time, requiring constant attention until July, when, his male classmates acting as pallbearers, he was buried. It was my second contact within a year with death of an acquaintance. Paul Clem's death had occurred earlier, and left me the responsibilities of the Mason Garage Shop. Father had hired another mechanic, a second generation son of a Czech immigrant family, who was regarded as a man of mystery in WaKeeney because he had worked in Wichita as a production mechanic in an aircraft factory.

When I returned from Bingham's funeral, our mechanic was standing near a venerable Cadillac, stalled near the gasoline pump at the curb in front of the garage. Our mechanic looked puzzled, and he was. The Cadillac had stopped, filled its tank with gasoline and refused to start. There was spark, there was gasoline in the carburetor, the starter cranked vigorously, sometimes there was a faint whiff of smoke, and that was all.

When I had changed into coveralls, I made rough checks of ignition timing and valve timing, and both were far enough from right to excuse the iron horse for balking. The engine was out of time at the camshaft, and the next thing to do was to get access to the timing chain and gears. This our mechanic could do, and sure enough, he found the short timing chain and sprockets worn out, with the timing marks on crankshaft and camshaft sprockets far from their proper alignment. We would have to order parts from Kansas City, and the Cadillac owner and his family would have to stay at the comfortable Staatz Hotel, across the alley from the garage.

Getting to know the owner of the ailing Cadillac is one of the experiences that makes me happy that I did not, like some of my classmates in high school, go immediately to college. Doctor Coil, the Cadillac owner, was a surgeon, on vacation with his wife and son from a practice in Mexico, Missouri, about a hundred miles east of Kansas City. His attitude toward people and the world in general was one of confident acceptance. He and his family would be quite comfortable, he said. They were on vacation, and WaKeeney would be the place to spend part of it. His wife and son would be content to read, sitting under trees in our Court House Park; they liked the food at the Staatz Hotel. He, himself, would be reasonably content, but—was there any place near to go fishing? Father, not a fisherman, knew the places where fish were caught, and he arranged for me to take Doctor Coil to one of them, a deep pool in a creek that ran into the Saline River about ten miles north of town. We would go in our Packard touring car, and when Doctor Coil saw my brother, then about eight years old, standing by wistfully, he promptly included him in the party. I drove through a pasture to stop the car near the pond. We assembled bamboo poles, lines, cork bobbers, sinkers, hooks and grasshopper bait. No fish came to our tackle. The evening light faded, and brother Howard began to be restless. Doctor Coil gathered Howard into his lap and started telling him stories. I put tackle into the car, started the engine and attempted to drive up the road that led to the pond. It was no use. We were hopelessly stuck. The eighteen-year-old who could diagnose a slipped timing chain had not had sense enough to park on solid ground. It was not far to a farm house; I called Father, and he came and brought us home. Through all this, I heard not one word of reproach from either Doctor Coil or Father. Reality was punishment enough. Our mechanic, faced with a straightforward assembly job, fixed the Cadillac quickly, and it ran smoothly. Later we received a postcard from Doctor Coil, from a mountain town in Colorado, reporting that the car "runs like a rubber-tired buggy."

WORKING MACHINES

To deliver smaller farm machines to customers, and to transport me, the high school student, to attend out-of-town athletic events, on rare "dates" and on troubleshooting trips to the countryside, we used automobiles and light trucks that had come to us in trade, often as down payment for new machines. Two of these exemplified what might have been called the Mason Fleet. One was a 1923 Packard Six touring car, the other a Haynes Six of about the same year model. Both served for several years.

We probably acquired the Packard sometime in 1927. That was the time when the Union Pacific Highway was being graded up to a level above the surrounding countryside by a then-new type of grader, which scooped out deep and wide gutters on each side of the roadway, throwing the yellow clay that underlay the loam that, thirty years earlier, had supported prairie. The plan was to coat this new surface with gravel, which would be moved back and forth across the roadway to maintain an all-weather surface. Sandy gravel was obtained from washes near creeks and hauled to the new roadway in fleets of Model T Ford trucks. These had been modified by having their drive shafts shortened so that an auxiliary transmission, called a Warford Tee, could be inserted behind the regular Ford engine-and-transmission unit to provide increased torque. The drivers of these trucks formed loose crews of individual entrepreneurs and were paid by the cubic yard of material delivered. Such an arrangement freed the road contractors from the need to own fleets of trucks and, more important, from the expensive and frustrating task of keeping the trucks on the job. The owner-drivers, optimistic young men, would haul gravel for four or five days, then bring the trucks to the garage to have the connecting rods adjusted over the weekend. I grew accustomed to spending weekends looking at the lower ends of Model T engines. The older engines had short pans that gave easy access to the first three connecting rod bearings. The fourth rod could be adjusted, but required a special wrench, and offered opportunity to drop bearing nuts into the oil pan. To recover them then, you had to remove the entire engine. (The same held for their cousins, the transmission-band nuts.) Since the fourth rod got a better chance at lubrication in engines frequently allowed to run short of oil or subjected to long uphill pulls, the fourth rod tended to need less adjustment, and mechanics often neglected it. One reason our garage got gravel truck business was that, being country yokels driven by conscience, we always adjusted all four bearings. This was good for business and conscience, but it was hard on knuckles.

Travelers, as well as gravel truck operators, tended to be optimistic. As the number of all-weather miles grew larger, more and more motorists would start from Kansas City, assured of about two hundred miles of smooth riding and hoping for good weather beyond Hays or Ellis, Kansas. When their luck played out, they were likely to store their cars in our garage and take the train, either back East, or on to Denver, picking the car up on the return journey. If the car was of marginal value, they might simply leave it garaged and never return to claim it. In this way we became owners of a Packard Six. Having stored it for a year, then advertised it in the local newspaper, Father could sell it to himself, at auction, and thus have legal title.

Clem, our mechanic, looked it over and pronounced it worth rehabilitation. He had persuaded Father to buy paint-spraying and body-bumping tools, and the Packard got a thorough restoration. In Plains Gray Acme Proxlyn, a lacquer competitor of Dupont Duco, it was made to shine. A few months later it became an early beneficiary of our new Storm Reboring Bar and the small lathe with which Clem turned pistons down to size. His finish on pistons was far from high polish, but they wore well when broken in discreetly. The Packard became a moving advertisement, and I, as a high school junior and senior, was eager to drive it here and there, often with several of my male friends and their dates in the rear. The body of the five-passenger Six was identical to that of the seven-passenger model, except for the jump seats provided for the latter, so that young bodies had ample room to intertwine, in patterns I, as driver, was only dimly aware of. Another attractive feature of the car was that the exhaust pipe could be sprung down from the manifold and pushed to the side, so that either a quiet or a noisy car was available with a minor adjustment.

Generally speaking, the Packard was designed conservatively. It had two features, however, that were not to be passed on to later generations. The first of these was the fuelizer. The carburetor had a small chamber, interconnected with the exhaust manifold, in which a small amount of gas-air mixture and warm exhaust was exposed to a special spark plug, activated by the ungrounded end of the engine's ignition coil. The idea was to provide a small warm stream of burning fuel to blend with the output of the carburetor proper and thus make the engine run more smoothly at speeds just above idling. It was possible to adjust the device so that it seemed to work, but within a few minutes or a few miles, the special spark plug would foul with carbon, and whatever input came from the Fuelizer was only a poorly mixed addition to what the regular part of the carburetor furnished. Like most of those who "owned one" we soon cut off the fuel supply for the fuelizer, and lived happily with the result. The car would run seventy-five miles per hour on the level, or even slightly uphill, and at any sensible speed on the roads we had, it would go fifteen miles on a gallon of gasoline—a fair compromise between the customer and Standard Oil of Indiana.

The second special feature of this model Packard was the clutch. Unlike those in later-model cars, it was of the multiple-plate dry type. (The Hudson at this time employed a multiple-plate wet clutch.) The idea behind the elaborate engineering involved in these units seemed to be to make the effort of depressing the clutch pedal as slight as possible. With many surfaces to carry the engine's power to the transmission main shaft, lower spring forces were needed, and the leverage required at the pedal could be much less. Probably this was an engineering attempt to encourage women to drive cars, they being considered weak frivolous creatures by engineers and salesmen. When a car was new, the multiple-plate clutches worked very well, though they made the cost of the vehicles considerably greater. It was to be a few years before cost, in nickels and dimes, would be of great importance to designers.

As a car aged, the splined spiders on which the several clutch plates moved became notched in the plane of the plates so that, when engaged, the plates rode into the wear notches and had to be hoisted out of these grooves when the clutch was released to shift gears. There were many of these grooves, and, short of installing new spiders when a clutch was relined, there was little that could be done to make the units work better once groove wear had occurred. There was a way, however, to improve matters. The thickness of the plate linings could be increased slightly, even to the point where a few plates had to be discarded, whereupon a new set of unworn surfaces on the splines of the spiders could be exposed to wear.

The message to me from this clutch was, *make it simple.* In most matters, Packard had done this, but I found, in servicing later models, that their engineers were prone to carry along at least one untested innovation in each major model change. The most disastrous of these was the Skinner Oil Rectifier. This device used intake manifold vacuum to suck off excess oil from the lower parts of the piston skirts as they came to the end of their strokes, carry this oil up to a stove resting on the exhaust manifold, and there redistill the oil, ridding it of dilution from the over-rich mixtures of fuel needed for starting cold engines. As a mechanic, our man Clem thought, and I agreed, that the engineers viewed the rectifier as a way to control engine oil consumption. It would be a few years before arrays of piston rings were designed so that a good balance between cylinder lubrication and low oil consumption could be achieved, aided by better ways to ventilate crankcases and reduce vapor pressure at the bottoms of cylinders. At any rate, the Skinner device did not work well. The first announcement of its failure was likely to be a stuck piston, where too much lubrication had been sucked away, leaving a dry area where piston and cylinder rapidly reached welding temperature. The mechanic's cure was to cut off suction to the rectifier. The rectifier found better use on the Willys Knight Six, which employed two moving sleeves, instead of poppet valves, to move gas mixture and exhaust through each cylinder. With so many surfaces to be oiled and to waste oil, it became worthwhile to work out the fine points—to adapt the device to the engine.

By the time we acquired the Haynes, its manufacture had probably been discontinued. It had been towed into the garage next door, the one under the hotel, and was considered a basket case. Clem often visited Harry Staatz, the proprietor of the other garage, and he was attracted to the Haynes as something he could repair, and that, as he thought, Staatz could not.

The huge six-cylinder engine, with three and five-eighths inch bore and five inch stroke, was larger than the Packard Six engine and thus somewhat more powerful. Clem could patch the crankcase with sheet metal and gaskets and have a powerful chassis on which a box could be mounted to make it into a ton-and-a-half truck. The particular attraction the job held for Clem was that it would give him a chance to show off the Storm Reboring Bar.

As I remember it, Clem traded Staatz a good-looking Ford Model A roadster for the Haynes. Father was not amused but did not want to scold Clem. Staatz promptly decorated the Model A, and paraded it around, doubtless telling people he had bested Clem. Clem went about reviving the Haynes. He ordered a smooth sleeve, about five-sixteenths of an inch thick, from the Lee Hardware Company, straightened out a damaged connecting rod, measured the crank throw on which it had been running and had it rebabbited by the Watkins Company in Wichita. Everything went as planned. The block was bored out, and the sleeve, coated with white lead, pulled into place without incident. The hole in the crankcase was repaired easily, and the body of the car, behind the dash, was taken off and scrapped. Father, who prided himself on construction of wagon boxes, bound oak timbers around the frame to serve as bolsters for a truck bed. Here it was necessary to guess how much the rear springs would settle when the truck was loaded. Clem left the construction of the truck bed and the seat for the driver up to Father and me. I thought the floor of the truck box should slope toward the front, remembering with pain an incident in which I had started from a farm to town with a small pile of grain in an old truck and arrived without any of the grain, the rear endgate having sprung open, and the grain jolted toward the back and out. Father respected my guesses, so the box was canted smartly up at the rear, and the front endgate slanted up over the seat cushion on which the driver was to sit.

We had a second mechanic named Bill Herr, who drove the truck for us, a tall man who had no trouble reaching the pedals or steering wheel, but who had to assume a forward stoop in order to sit behind the wheel. We used the Haynes for a long time. It seemed to work effortlessly, and since we had never connected the speedometer, Herr was prone to underestimate its speed. On one occasion, he took a piece of John Deere equipment to the dealer in Quinter, twenty miles away, unloaded it, loaded up a large grain drill and returned to our garage in just an hour. We could only guess at his average road speed, but it must have exceeded sixty miles an hour. Herr was probably happy not to sit on that seat any longer than necessary, but he said he didn't hurry; he said he just moseyed along.

Father had one bad experience in reinforcing truck boxes. During the time that the Union Pacific Highway was being graveled, he took on the dealership for Stewart trucks. A young fellow had bought one to use in hauling gravel, and seemed to like it, except that he was disturbed when the cab shifted on the frame as he brought loads of gravel up at an angle from the road gutter to the top of the grade. Father said he thought he could cure the trouble. He went to the lumber yard, got a long piece of four-by-six-inch oak and had his blacksmith friend make bolts with which to clamp it to the frame. Only one load was required to demonstrate that the truck designers had anticipated some squirming of the frame as wheels came up over obstructions, one at a time. There were two holes, a bit larger than four by six inches, in the cab of the truck, where the timber had obeyed the laws of physics once to the right, once to left, as the truck angled up from the gutter to the top of the grade with its first load of gravel.

Good Roads
The Golden Belt

Until the autumn of 1925, when we became garagemen,
I had been only dimly aware of bright yellow bands, painted at
chest height, on the poles that supported telephone lines running between the towns that owed their existence to the Kansas
City-to-Denver line of the Union Pacific Railroad. In fact, by
1925, these bright yellow bands were fading. They were the work
of the Golden Belt Association, a confederation of Commercial
Clubs, forerunners of Chambers of Commerce, organized to
promote business interests in towns such as Russell, Gorham,
Hays, Ellis, WaKeeney and on through Oakley to Sharon Springs,
Kansas, and the Colorado line. There were no doubt lobbies to
urge the Kansas legislature to make money available to build a
through highway. This came about by fits and starts which I'm
not equipped to chronicle in detail. It would have been strange,
however, if the move for an all-weather highway did not encoun-
ter resistance. Trucks were already seen as threats to railroad
revenues, passenger service was still regarded as profitable, and
the financiers who had promoted railroads and the settlement of
the area they ran through had long ago lost interest in railroad
operations. They had, however, been granted a broad right-of-

way for their tracks and associated telegraph and signal lines, and giving up a portion of it for a highway was nothing to be rushed into. The advantages to a highway of following the rail route were great. The line had been laid out to avoid steep grades and long fills, at great expense and after many controversies. Towns on the railroad prospered; those far from it faded. When we went into the garage business, the Golden Belt Association was in decline, and a new effort was made in the name of the Union Pacific Highway. Significantly, the white bands on roadside poles that replaced the fading yellow ones were on the roadside nearest the railroad and, I think, on the telegraph and signal poles within the railroad right-of-way. Again I defer to historians, but my guess is that the railroad company had found that what the state gave, it could also take away; the railroad must not only permit painting on its poles, it must do the much more expensive job of protecting grade crossings with signals and, in congested areas, with crossing gates. It would be many years before Interstate 70 straightened the highway, detoured the less prosperous towns along the railroad and aligned itself to avoid, rather than run through, the main streets of the others.

MUD AND THE AUTOMOBILE

The rail distance between Kansas City, Missouri, and Denver, Colorado, via the Kansas Pacific segment of the Union Pacific System, is 640 miles. WaKeeney, Kansas, is the town nearest the middle of this distance—the distances to terminal points of the rail line, marked on the east and west ends of the WaKeeney depot, were 321 miles and 319 miles, respectively. Thus, our garage business was positioned well to provide overnight storage and emergency repairs for the well-heeled citizens who drove their city-used Packard Straight Eights from Kansas City or points east to family vacations in the Colorado mountains. Most of the long-distance tourists stayed at Staatz's hotel. The highway ran up Russell Avenue, directly in front of both the hotel and the garage.

Our garage's chief claim to prestige was its being the Chrysler Agency, only a year after the Chrysler Six was introduced to begin the battle to replace the mid-priced Buick as the way for prosperous businessmen and farmers to show that they could afford something more expensive than the Model T Ford. Dealing in new and used cars was far from the day-to-day occupation at the garage, however. Selling and installing exhaust manifold heaters on Model T Fords gave us some business. Starting tourists' cars that had made the run from Kansas City but refused to start after a night on the street kept us alert to the new developments in the industry. Ignition points that had built up a knob on one side and a crater on the other would fire the spark plugs when the engine was hot but often would not start a cold engine. Many car models were not designed to be run more than a few miles at what are now moderate road speeds. Vacuum tanks, used to entice gasoline from the rear tank to a position above the carburetor, depended upon an engine working at considerably less than full throttle. When the road had been straightened enough to allow sustained operation at road speed without slowing down for railroad crossings or sharp turns, power might cut off, only to return when the driver released the throttle, restoring intake manifold vacuum and thus coaxing more gasoline from the rear tank. Many cars, particularly the impressive Packard Straight Eights, were not designed to stay cool in hot weather when driven more than a hundred miles or so. Replacing fan belts was a common task, but fatherly advice to stop frequently and enjoy the view might have been more effective. Eventually, the Kansas Highway Commission provided an incentive to stop and read bits of Kansas history, but this was not until sometime in the late 1920s or the 1930s. To folk who did not consider a park planted with trees and shrubs the essence of rustic beauty, there was probably nothing that would induce them to tarry between Ellsworth, Kansas, and Denver, Colorado, if their equipment would continue to function. We learned about regional accents, family dynamics, roadside diplomacy and the workings-out of the automotive industry's long-running experiments in which customers, rather than racing or proving grounds, provided answers to problems.

Much of the junior mechanic's working time was spent looking at the lower connecting rod bearings of Model T Fords, removing the bearing caps one at a time, filing the mating surfaces a given number of strokes with a mill file, replacing and tightening the caps, checking for proper fit and repeating the process. When too much stock was removed, thin brass or paper shims could be cut and placed on the bolts that held the bearing together. As in all crafts, speed and accuracy in the work depended upon the ability to make good initial guesses. Another frequent job was to adjust the transmission bands of the Model T. This was rather easy, but replacing bands that had worn out their lining was not. Replacement bands with one top lug that could be released—so that the band could be pulled out, relined and reinserted— soon appeared. In the bad old days, it had been necessary to remove the whole transmission cover to remove bands. When using the new-style bands that Ford soon adopted one had only to remove a nut and washer from the extension of each of the three pedals and place them safely away from the opening in the transmission case. Older mechanics usually stuffed rags down beside the drums to catch any part that might be dropped, either in running the nut off the pedal stub or replacing it once the relined band had been inserted and teased around into place. Younger men, of which I was one, bravely held on to the parts and didn't drop them. If one was dropped, it fell to the bottom of the oil sump, and after one attempted to retrieve it with a magnet at the end of a special tool, he usually removed the engine from the car, talked the owner into having a considerable overhaul job done "now that the engine was out" and tied up the car for a few days. I don't think I ever dropped a transmission band nut or washer, but I remember gritting my teeth in the effort to avoid doing it.

Far more time was spent putting on and taking off tire chains than in making mechanical adjustments to the cars themselves. If rain fell on Friday night, farmers would install tire chains at home, drive to town for Saturday shopping and, as often happened, find that roads had dried up before late afternoon. Then they might drive around to the garage to have the chains removed, washed and put in the backseat area of the car. The reverse of this often happened also; coming to town dry, the farmer would see rain in midafternoon and would drive around to have chains installed for the trip home. Often he would need a new set of chains—a matter of six or eight dollars. One brand of tire chains dominated the market—Weed—and one rural mail carrier, whose route was just over fifty miles long, cheerfully bought a new set of chains every four trips. Kansas was a dry state and WaKeeney was near the driest part, yet some February-to-May periods would see rain almost every day, and our drayman would dump off piles of bagged tire chains from Lee Hardware Company in Salina three or four times a week. I learned to put on chains so they would not come off; what was more important to me, I learned how it felt to be uncomfortable at work. Cold concrete, wet or dry, soon chilled my right shoulder and hip, and there might be little respite for an hour or two. It came as a welcome revelation that, after the first pair of chains, nothing hurt, and there was a fierce joy in being on top of the job.

THE JUSTUS BROTHERS
GO COURTING

Through 1925 and well into 1927, the Union Pacific Highway wriggled westward across Kansas in projects of ten to twenty miles' length. Travelers became accustomed to driving on completed segments, off onto detours, back onto graded-but-not-surfaced segments and finally to their destinations. Along the way, as the road materialized mile by mile, there were opportunities to observe the behavior of men at work, the capabilities and weaknesses of machines and the layered structure of the soil, down to a depth of three feet or more, as powerful grading equipment buried the Kansas loam under a foot to a yard of the subsoil that had been holding it up. The subsoil varied from place to place, but particularly in the western part of Trego County, it was a yellow clay, dusty when dry and unbelievably sticky when wet.

I have written of the exertions of the gravel trucks. The off-duty behavior of truck drivers occurred outside my ken, though absences due to hangovers and banter among drivers bringing in trucks for repair suggested that life on the job was far from serene. I did get to learn something of the way in which the roadbuilding affected the social behavior of the young. For what seemed to be an eternity, the highway west from WaKeeney remained graded to full height but not surfaced with gravel. The dances WaKeeney youth attended took place at Voda, seven miles west, or at Collyer, seven miles beyond Voda.

None of what follows would have happened if WaKeeney's Presbyterians had remained content with their old church. They were the upper crust of local society, however, and they decided to build a fine new church. Their architect recommended Justus and Sons as builders, and the elder Justus, with his two sons, moved into rented rooms and, over a year or more, put up an impressive building. Justus' sons were husky, good-natured and skillful; they were at home with brick, stone or wood. They were also single and lonely. On a Saturday night in the springtime, they got their Durant out of the garage and, dressed in their finest clothes, escorted Mrs. Bell's daughters to a dance at Collyer.

Sometime around midnight, I, who was taking my turn sleeping at the garage to provide twenty-four-hour service, was roused by young men who drove by the garage, but I was not fully awakened. I drifted back to sleep. Later, at perhaps two o'clock, more mature messengers roused me, not being content merely to deliver a message, they made sure that I got out of bed, stood up and heard clearly that the Justus boys and their dates were stranded, their car not running, about six miles west of town, far from the nearest phone.

We had a 1922 Dodge touring car that we used as a mud car. We kept heavy-duty tire chains on it; the rear fenders had long since been worn and torn away. There was no top. On the floor beside the front seat, we kept an assortment of spades, shovels and leaves from automobile springs with which to deal with mud. I got into my rubber boots, put on a slicker and rain hat and started west. Rain was falling steadily, and the Dodge, its chains clanking, carried me without incident to the defunct Durant and its occupants. The four young got into my backseat, covered themselves with a blanket and were not heard from further. A sincere effort had been made to return the Bell sisters to their mother promptly, and, if they did not turn up pregnant in due course, their reputations were saved. I was, of course, only an honest broker in these affairs.

No sooner had I turned the Dodge toward WaKeeney than the rain stopped. Within half a mile, the Dodge's right front wheel had grown to the diameter of a wash tub, with sticky clay mud fouling the space between wheel and chassis until the car could not be steered and would not proceed in high gear. The Dodge then threatened to stall in second gear and finally was immobile even in low, which had a reputation for being unstoppable. I got out and dug with a spade with spring leaves and with my hands until the wheels—the right front was only the worst of four—would roll. We proceeded through first gear, second and perhaps a short way in high, then the wheels fouled. It was back to the mud tools. This happened again and again. No calls for help, no protests and no offers of help came from under the blanket over the back seat area. I assumed that all was satisfactory. We arrived at the Bell residence, on WaKeeney's paved streets, in time to see good Christians getting into their cars to go to church. The girls were home, the Justus boys were on the porch to explain, the Dodge was soon on the wash rack, awaiting the high pressure hose and I was on the way home for breakfast. One of the Justus boys may have married one of the Bell girls, but there was certainly no hurry. The Presbyterian church was completed. Local builders waited in vain for it to show some structural weakness. Years later, there was not the slightest crack in the plaster nor gap in the mortar between bricks.

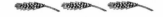

MUD FINALE

It must have been the fall rainy season of 1928. Paul Clem had converted the Packard touring car to a wrecker to replace the aging Dodge mud car. The converted Packard had an I-beam boom and space for spades and other tools where the backseat had been. Clem had soon died and left me, at twenty, in charge of mechanical work and road service. The highway west to Collyer and beyond had been graveled, the Warford Tee modified Model T trucks that had hauled gravel were gone. On a summer afternoon after a rain, we received a call to tow a car up onto the highway grade. It had veered across the road to the left shoulder, the south one, since it was headed west, at a high culvert and draw, near the Round Barn. This barn, which was also south of the road on the west side of the draw, was a landmark. The huge structure testified to its owner's identification with farming and, as I was to learn, his love of draft horses.

The distressed automobile was deep green in color, an open roadster of a foreign make. I was too much taken up with getting it back onto the road to be curious about its make and model, but it was huge. Attempts to back it onto the road had left its rear wheels free of the ground, its ample belly firmly wedded to the mud. It looked like an ideal job for the I-beam crane, and

I soon got a chain from the crane beam to the rear axle of the car, placing the wrecker as far over the road shoulder as possible without risking the possibility that the wrecker might join the unfortunate green monster on the sticky clay slope. It was, I recognized, the same stuff that, a year or more earlier, had trapped the young Justus men and their Saturday night dates.

It took only a few minutes to prove that the Packard could not free the car and restore it to the road. The driver and passenger joined me in the wrecker, and we went back to WaKeeney. One of the men was tall, large-framed and thin; the other was shorter and a bit more plump but not obese. Both were of middle age, and both spoke with what I took to be New York accents. They were, of course, not cheerful, but seemed to take trouble in stride and regard whatever happened as an unfolding life in which eventual success was taken for granted.

The next day we labored mightily, and a considerable gallery of farmers gathered. The sun warmed the scene. Father dug mud from in front of the mired car and caused the only protest from its owners. He was about to bang the edge of his spade against the car's front bumper when one of the owners shouted, "Oh, please! It looks awful now, but it's a beautiful car. Don't bash it." Father was squelched. Gradually, a plan for rescuing the car emerged among the onlookers. Chains, cables and ropes, perhaps a hundred feet in length, were strung from the front axle of the car, out across the wide gully bridged by the culvert and tied to a strong pier that supported the culvert on its west side. When several men spaced themselves along this cable and pulled sidewise in unison, the front of the car moved, not far enough to free it, but enough to show that it could be sprung from its muddy bed. Father had disappeared. Soon I saw him on the west side of the draw near the Round Barn. With him was a young farmer and a four-horse team, trailing doubletrees and a connecting beam. Hands untied the long cable from its anchor on the culvert pier and attached it to the horses' pulling beam. The farmer spoke to the team, and the green car moved gently down the slope, across the draw and up on the other side. The owners of the car approached Father to pay him, but he refused pay, introducing them to the driver of the team instead. In explanation, all Father would say was, "He helped you. We couldn't."

The owners did get to pay us a small amount, however. The car's wheels, fouled with mud, caused a severe vibration and, after driving slowly back to the garage, we cleared the wheels with the high-pressure hose. Father, who alternated between being the mature garage owner and the man with the spade, got to know the owners much better than I. He told me that they were en route to California, where they would be engaged in composing and producing a musical show. I have wondered since if they might have been Rodgers and Hammerstein, Jerome Kern and someone else or Rodgers and Hart. Apparently the show they were incubating could benefit from the experience of crossing the country by car, learning the rural culture firsthand. Whatever the case, what they experienced near WaKeeney would not have inspired *Surrey with the Fringe on Top.*

CRANKING UP

As Kansas highways became straighter and smoother, the "big" cars began to starve their engines for gasoline, even when there was nothing wrong with them mechanically. Fords, Model T or Model A, did not have this trouble. The difference was that, in the Fords, the fuel tanks were higher than the carburetors, so fuel could run downhill all the way until the tanks were indeed "out of gas." (There were one or two much-touted exceptions. When a Model T was low on gas and a steep hill had to be climbed, it was sometimes necessary to back up the hill to keep the fuel in the tank running downhill.)

The cars other than Fords, in the 1920s, had their fuel tanks at the rear, and fuel was sucked into a dash-mounted vacuum tank by the reduced air pressure in the engine's intake manifold. As long as roads were crooked and rough, livestock had to be avoided frequently, and city traffic required frequent stops, the vacuum tanks were adequate; but when the roads were straightened and smoothed and major highways were fenced, high-speed travel, requiring constant high-power output, was possible, intake manifold vacuum became marginal or disappeared, and there was no force to bring the fuel forward from the rear tank. After the quart of fuel provided by the vacuum tank's lower chamber was exhausted, the car sputtered and started to die. If the driver eased up on the accelerator, it might "catch" as fuel was again drawn from the rear tank, only to die again shortly after highway speed was resumed. Motorists and garagemen had to suffer from this trouble until about 1926 or 1927, when engines were equipped with camshaft-operated fuel pumps. Then better highway performance opened the way to problems with engine valves and bearings and cooling system deficiencies. The automobile industry seems never to escape the increasing demands better roads allow drivers to impose on machines.

Hart-Parr Rescue

The good times at the garage were sandwiched between 1925 and 1930. When Father took on the agency for John Deere farm machinery, in 1927 or 1928, the Deere salesman's pitch centered on the Model D tractor, a much reworked version of one of the early successful medium-sized farm tractors, the Waterloo Boy. Neither the Waterloo Boy nor the Model D had the two-cylinder market to itself, however. The Rumley Oil-Pull and the Hart-Parr were competitors. The Rumley used a smaller version of the large engines used to power threshing machines. It featured a smokestack-like exhaust, and I think it used kerosene as a coolant as well as a fuel. It never achieved wide popularity in our area, though we did trade for one. Hart-Parr machines were likewise not popular, though their Kansas City distributor sold them aggressively. At one point Father accepted the advertising literature for them, but he did not stock the tractors or parts for them. Though a good machine, the Hart-Parr had a clumsy appearance, and I was happy not to be a Hart-Parr mechanic, though I was once pressed into service as one.

A Czech immigrant community with its first-generation offspring was centered around Voda. In general, they were prosperous, friendly and able to speak English without more accent than any of us. Many of them were good mechanics, and their farmsteads were usually better-kept, more often painted and more likely to have vegetable gardens than most.

It was from one of these farms that I received a call to help a group of farmers who had overhauled a Hart-Parr tractor and were unable to get it to start. When I reached the farmyard where they were working, they had belted the Hart-Parr up to a John Deere so that it could be cranked by belt-power instead of muscle, and they were in process of disengaging the camshaft timing gear to check whether or not it was engaged to the crankshaft according to marks usually placed on these gears. As I remember it, there were no such marks, but it is fairly simple to determine correct meshing of the gears. The rule is that, when one of the cylinders has both valves closed at the top of its compression stroke, the other cylinder, turning on until it is at top-center, should have both valves barely open—the intake opening and the exhaust closing. This verified, it was apparent that valve timing was not the cause of the trouble. With the gearing buttoned up, the men turned the engine by hand until I could check this out and also determine that the magneto delivered a good spark to the cylinder that was ready for it.

Then the men started the John Deere, opened the priming cups on the Hart-Parr—a feature it shared with the John Deere—and proceeded to crank away by belt power. The first thing I noticed was that, while the priming cups hissed as the pistons traveled up and down (actually back and forth) in the cylinders, they seemed only to gasp, not to draw in a full charge on the intake stroke and expel it with vigor on the following compression stroke. Pondering this, I noticed just one gasket profile that did not look quite like a factory job. I asked whether any of the gaskets had been handmade. I hesitated to ask any questions about the work that had been done, because it all bore the mark of expert and careful attention. There were no ragged edges. But one gasket had been handmade from a thick heat-resistant material. It was between the case that held the valve-rocker arms and the cylinder head; its outer lower surfaces kept the rocker-arm case from leaking oil. Exhaust flanges sprouted from each side of it, in front, and connected to an exhaust manifold. The carburetor did not connect to this gasket, but was nearly twelve inches directly behind it, on the rocker-arm case. Seized by inspiration, I disconnected the air cleaner and put the palm of my hand at the intake of the carburetor. When the engine was cranked, there was no suction. It was apparent, though it could not be seen, that the homemade gasket should have had a hole centered between the two holes to handle exhaust. This had not been made. It was a small matter to remove one of the long valve push rods, which had large ball-like ends, and jam it through where the hole should have been cut. With good luck, the mating surfaces would hold, and the gasket would not be ruined. When a breathing hole for the carburetor had thus been made and parts replaced, the Hart-Parr burst into song, and all of us were ready to weep for happiness. I did not question the men as to whether one of them, or someone at a shop in town, had made the one-hole-short gasket. I cautioned them that they must disassemble the parts held apart by the gasket and make a clean, full-sized hole. This was scarcely necessary—they knew their business well enough. What I have wondered about in later years is whether or not the man who cut the gasket also knew what he was doing and was, in fact, pulling the mechanic's equivalent of the summer camp trick of looping up the top sheet of a tenderfoot's bed (short sheeting)—making a shallow sack where a through bore is needed. When the men offered to pay, I suggested they wait for a bill from my father, and I'm sure he would not have sent one.

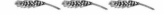

The Ground Floor

In the 1930s, the world was between wars, and the Depression and dust storms had come to Kansas. Magazines such as *Popular Mechanics* began inviting untutored young males to "get in on the ground floor in aviation." A man named Heath had built an airplane, powered by a four-cylinder Henderson motorcycle engine that would carry him aloft. He had named the airplane the Heath Parasol and set up a company that sold plans and kits with which a patient mechanic could build the plane. An urge to be on the ground floor was not the only reason to attempt a building project. Escape from boredom was probably a stronger motivation for most of the young men who started airplane projects. Work at the garage consisted mainly of selling small amounts of gasoline and oil and tidying the place. I read classified advertisements in mechanics' magazines over and over in the hope that I could assemble an airplane from large pieces, rather than build it from scratch. Eventually, two aircraft building projects were under way in WaKeeney. Down on the bypass (the highway had been moved to the south side of town), Joe Masopoust started putting together a Heath Parasol, and I, up on Russell Avenue, in the small corrugated metal building I was using while the new garage was being built, bought, with money provided by Mary, a plywood and glue fuselage and landing gear and a pair of wings with ailerons, each fifteen feet by five feet. Henry Erichs' truck put the things off at my shop. With the wings resting on their noses along the west wall of the shop, the varnished wood fusilage resting on the axles of the landing gear, and everything visible from the street, I was publicly committed to the project. I was excited, but not hopeful. Down deep, I was wondering just when and how the whole project would collapse. Meanwhile, I would be occupied.

Masopoust was the better workman. None of the construction problems were beyond him except construction of an engine or money enough to buy a Henderson motorcycle engine in working condition. Eventually Masopoust's plane got off the ground, if only briefly. By grace of providence, my effort fizzled before it could pose substantial danger to life and health.

I set to work to modify a Model T engine for use in an airplane. The flywheel end would be forward and the usual cranking-up end would be buried just ahead of the pilot's feet. On the ground, the flywheel served as the main part of the magneto and the means to distribute oil to engine parts. The flywheel picked up oil from the crankcase and dropped it into a funnel-shaped piece. The oil ran from there through a tube to the timing-gear end of the engine, returning over scooped-out depressions in which connecting-rod bearings splashed to oil themselves and the pistons above them. In an airplane application, an oil pump was needed. It could be installed where the Model T's generator had been, pick up oil from whatever part of the original crankcase sump was left after its big bulge had been trimmed to fit the nose of the airplane, and push the oil into a copper pipe flattened to a nozzle at the end and containing strategically placed leaks to fill the pans under the connecting rods. If I did not get it right the first time, it was a small matter. The garageman's way in such cases was, and may still be, to fill the engine with enough oil so that all the connecting rod bearings would splash. Excess gray smoke would indicate excess lubrication, and fine-tuning could be postponed.

I had less trouble providing ignition to the engine. I built a shelf across the end of the oil pan where the fan pulley would have been, found a tractor magneto that turned clockwise from the drive end and had an impulse-starting mechanism that could be inactivated. The impulse starter needs a bit of explanation. Tractors, at least at that time, were started with hand cranks, and it was nearly impossible to spin the engines fast enough to produce a long spark for starting. Instead, an impulse coupling allowed the magneto to lag almost a half-turn behind its proper position, while cranking the engine wound up a strong clock spring. As the engine passed firing position, a catch would release the held-back magneto, the spring would propel it smartly forward, and a fat spark would occur at the spark plug. Aviators frowned on the clock spring that made an impulse starter work, insisting, wisely, that the magneto be solidly coupled to the engine. Jittery passengers are inclined to wonder whether it is safe to fly in an airplane that takes a lot of cranking before it starts. Lack of an impulse-coupled magneto is probably the reason. A prospective passenger is entitled to worry, but a slow-starting engine should not be cause for alarm.

To test the engine, I bolted a Model T flywheel to the crankshaft, inserted a nail through a hole in the flywheel and found a piece of aircraft control cable with loops in the ends. I could wind the cable around a ridge on the outside of the flywheel, connect its inner loop to the nail and pull the engine through its cycle by yanking on the cable. Gasoline in the carburetor would ready the rig for testing. Still I hesitated. I had heard that, unless exhaust valves were protected from direct contact with cold air, they would warp and burn. I wanted a little time to get my nerve up, so I constructed four exhaust stacks to fit not quite tightly against the cylinder block and to stick up at an angle that would give a properly jaunty appearance. I had already done some bending and welding on the pieces that were to be the stabilizer, elevators and rudder. A pint-sized can had been attached by means of a coupling to the fuel line to the carburetor. There was no one in sight on the street. I gave the cable a pull. Success came on about the third or fourth pull. The choke had been set slightly, the throttle had been opened a bit, and the spark advanced as far as I thought safe. When the engine caught, the nail I had inserted in the flywheel to hold the starting cable flew past my head. A staccato set of pops came from the exhaust stacks and blended into a rough roar when I opened the throttle slightly. After less than a minute, I shut the engine down. There was no radiator yet, and any extended testing would have been sure to overheat the engine. Almost as I felt the sweat of excitement cool my back and run down from my armpits, I knew that the airplane project was dead. I may have run out of oxygen or acetylene or both and would certainly have little money for more. The inescapable intuitive truth was that I could not complete a large project. I was temperamentally unsuited for what I was later to learn was the trampling underfoot of the meek on the ground floor.

The fuselage with its engine still sat in the shop when, in January 1934, I enrolled in college. The wings were eventually stored up under the gambrel roof of the big barn at the farm. A few years ago Howard told me that he and a friend had made runners for the fuselage from discarded automobile bumpers and used it for a sled. I think they towed it. Howard had better sense than to put much work into such an unlikely project as a Model T air sled.

While I was working on my Model T power plant, Masopoust, down on the bypass (it has since become the main stem) had accumulated kit materials for a Heath Parasol airframe, controls, landing gear and wings. All he lacked was power plant and propeller. These he supplied by responding to an ad in a mechanics magazine that offered a new, complete, Le Rhone engine and propeller in its original case. I remember that these units could be bought for fifty to seventy-five dollars plus freight. For Masopoust, who was essentially building the airplane from scratch, making the many wing parts and gluing up fuselage sections came first, and rightly so. If one really enjoyed hand work, he could buckle down to the long process of gluing up ribs, placing them on the spar pieces provided and covering and doping them, and prove to himself whether or not he enjoyed the work enough to see the project through.

Masopoust knew that he would have to shift the wing forward from its usual position to make its center of pressure coincide with the fore-and-aft center of gravity of the loaded airplane. What bothered him most was welding joints for fittings and tube structures for the undercarriage. He asked me to weld a few of these, and I was flattered. I had seen some of his welding and was sure it was better than mine. Nevertheless, by bearing

167

down, I could and did weld several joints and convinced Masopoust that he need not fear for the strength of his own work. So far as I know, none of the welded joints on the plane failed when it was finally tested, long after I had undergone the metamorphosis from country mechanic to scholar.

The engine that carried Lindbergh from New York to Paris in 1927 was a Pratt and Whitney radial. Seven cylinders, resembling those on a motorcycle, stuck out like spokes of a wheel from a central crankcase—that is, they were arranged radially. To an unpracticed eye, the Le Rhone did not look a great deal different. Machined and polished to military specifications and nestled in its shipping box, it had a sculptural beauty. The crucial difference between it and the radial engines manufactured by Pratt and Whitney or Wright was that the Le Rhone, in addition to being a radial, was also a rotary. This means that instead of remaining steady in place on the engine mount as did the cylinders of the ordinary radial, those of the Le Rhone whirled around when the engine ran; the crankshaft, instead of whirling the propeller, anchored the engine to the airplane and, itself, remained a fixed part of the aircraft. To a country mechanic short of money this difference seemed to be something that engineers had coped with, a choice that had produced usable power plants of both types. An astute buyer might have raised the question: Why is this beautiful and sophisticated engine available for only fifty dollars? Masopoust probably assumed that fashion's whim was the major reason for preferring the fixed-cylinder radial to the rotary. He exercised a trust in anything that could be called The Factory, much as the sucker at a sideshow involuntarily gave his trust to the well-dressed and self-assured barker. I would have, too, had I had fifty dollars. Barnum was only partly right. A sucker is born of circumstances, not just once a minute, and not of selective genetics, but of the universal thrust to be significant. To be a significant airplane builder, one must have an engine; fifty dollars plus freight was the going minimum price.

What Masopoust and I did not know, and the engineers of Le Rhone should have known, was that any rotating mass is a gyroscope, stubborn about changing the direction in which its axle is pointed; the greater the mass, the greater the stubborn persistence in direction. To make a maneuverable airplane, the rotating crankshaft and propeller must provide less of the un-desired gyroscope effect than would a rotating propeller that took seven cylinders and their associated parts along with it.

To country mechanics like Masopoust and me, the prob-lem presented in substituting the Le Rhone engine for the specified Henderson motorcycle engine was a matter of proper engine mount fittings and modifying the struts that fixed the wing to the body of the aircraft. The wing would have to be moved forward from its intended position, so that the designated center of pressure for the wing—about one third of the wing's chord or width back from its leading edge—would be directly over the center of gravity of the airplane when loaded for flight. Masopoust could accomplish this; fine-tuning of the fore-and-aft trim could be achieved by adding or subtracting small amounts of weight at the tail skid which, being far to the rear, had a great deal of leverage against the forward parts: propeller, engine, tanks, landing gear and so forth. Masopoust, or any competent mechanic, could cope with the center-of-gravity problem. I have always regretted not seeing his finished airplane. The completed and doped wing was a beautiful piece of work, a foretaste of the much smaller-scale wings my son Peter made as a high school student many years later for model gliders I towed into the air for him while he held their tethers from the rear seat of our Volkswagen.

Howard, who stayed in WaKeeney for a few years after I left, told me that Masopoust's plane was eventually tested. An airplane pilot from Oakley, west of WaKeeney, inspected Joe's work and found it good, but he pointed out that the large powerful engine would require redesign of the tail parts of the Heath Parasol. He indicated the changes needed, and Masopoust constructed the larger, stronger tail. Masopoust himself probably piloted the plane for taxi tests, running down a grass runway without lift-off. It is possible that Masopoust's friend from Oakley came down, and it is also possible that Masopoust, having gained experience in taxi tests, gunned the engine and allowed his plane to take off. At any rate, Howard reported that the skyline could be seen under the plane—it was airborne—before it crashed. Whether it came down on one of Masopoust's beautiful wingtips or its nose I do not know, but it was, after a great deal of work, a thing of the past.

I saw Masopoust years later, in 1947, when our family, heading west to a job at Whitman College in Walla Walla, Washington, stopped to rest in WaKeeney at Mary's and brother-in-law Ray Musgrave's house. We were transporting bag and baggage, furniture and everything, from a temporary post-World War II job at Purdue in an ancient Chevy school bus. A rear axle shaft had been replaced in Cameron, Missouri, and I wanted to be sure that the brake drum was secure on its taper. Masopoust had the right tools, and he secured and keyed the nut. He had taken over his father's shop and filling station and, I'm sure, sold gasoline that would burn. We talked of families and weather. We did not speak of airplanes.

HANGING ON

COMBINING WHEAT

To say that Father, overtaken by dust storms and the post-1929 Depression, went broke in the garage and implement business is too simple. Father was far from either a wild speculator or a vulnerable one-enterprise merchant. As business at the garage fell off, then hovered around zero, he floated one scheme after another in attempts to stay solvent, at least to the extent that his and mother's home farm not be foreclosed from under them. It had been mortgaged to finance the garage venture, and he exerted himself to protect it.

One of the early ventures was to take a twelve-foot #5 John Deere combine out of stock, hitch it to a John Deere Model D trade-in (combines were not self-propelled in those days) and cut wheat for himself, his lawyer friend, William Wagner and anyone else who would pay $1.75 per acre and furnish trucks to haul the wheat from the machine. In many cases, he also furnished trucks, probably for an additional fee. No one expected this venture to turn a real profit. It would repay at least some of the invoice price of the combine, and keep me, my sister Mary (who transported field hands, lunches and drinking water to the fields) and Wagner's two college-age sons, Kenneth and William, Jr., my contemporaries and one-time high-school friends, employed for part of the summer.

The summer was dry and hot, ideal weather for harvesting wheat. We could start early without waiting for dew to dry and work late, as late as we could see to drive the rig, about nine o'clock in the evening. The Wagner sons did not think of themselves as mechanics; they were fledgling lawyers, but any doubts I had about their ability to run the rig faded quickly.

They grasped, without being told, that the way to make money was to keep the rig running. Consequently, whenever we stopped to unload wheat from combine to truck, they promptly manned grease guns and started wherever they had left off at the last stop, greased bearings along each side of the machine, noted where they had stopped greasing, mounted the machine and started cutting wheat, not stopping again until the grain bin was again ready to be unloaded. My duties were to check the machine and fill fuel tanks early each morning, spell the operators one at a time while they ate lunch and dinner, keep an eye on the mechanical condition of the outfit and do any repair work that came to the garage shop. Very little came.

As the Depression deepened and farmers found it less expensive to let their wheat stand than to cut it, business turned itself off like a faucet. We who kept the rig operating had only one bit of business-related intelligence to keep in mind. We should not cut any wheat that made less than seven bushels per acre. Our charge was $1.75 per acre, wheat was selling for twenty-five to twenty-seven cents a bushel, and we could not expect a farmer to pay our per-acre charge from his own more-than-empty pocket. Some of the acres we drove over put our judgment to the test. On a few occasions, we detoured hillsides and cut only the parts of fields where grain was heaviest.

Doubtless our elders found the season grim. Not so we. In ten days, we harvested five hundred acres. After working until dark and having an extra dinner, we gathered around Mary, the less-gifted pianist of my sisters, and wailed popular songs from sheet music until drowsiness suddenly appeared. I, at least, had a warm bath and slept soundly until four or five o'clock in the morning. I remember only that, in mid-afternoon after ten days and five hundred acres, I drove the rig into the back lot of the garage, got off it and went home. Only then did I feel exhausted. I must have slept for a dozen hours and waked still a bit dizzy and disoriented. There would still be fields to be measured, with the Wagner boys taking the lead in this work, and machines to be put away. I would stumble through this, but harvest was over.

Pigs

What I now remember about the Depression is that it lasted a long time. There was not just one summer of combining wheat that sold at four bushels for a dollar, there were several. The new #5 combine we used that first year was sold, and after that we used a #1—an older, heavier machine, taken in a trade. Moreover, the intelligent Wagner boys had to be replaced with less-eager hired hands. Father hatched a plan to get more than twenty-five cents a bushel for the wheat we raised. To finance the plan, he went across the street to the Trego County State Bank, to Charlie Hille, the bank's chief partner. This was fully approved by Father's garage partners at the WaKeeney State Bank. During the Depression, bankers liked to spread the risk.

At the home farm, father had, many years back, fenced about twelve acres with hog-holding mesh, topped by two strands of barbed wire. Now he planted barley for pasture in this pen. Very little repair would be needed. He would extend this fence to give entry to the stalls of the horse barn, which was no longer used, and convert the mangers in front of those stalls to self feeders, to be filled with dry ground wheat. To grind the wheat that he bought, he would take a ten-and-a-half-inch Letz feed mill out of stock, mount it to be fed from a truck, and power it with a traded-in Twin City tractor. So far, little new capital was needed. Hille would furnish money to buy about 125 thin but healthy stocker pigs, weighing forty to sixty pounds each. Then the pigs could begin turning twenty-five-cent wheat into pork. Hille found a trusted judge of livestock and provided him with a checkbook. The balance carried forward on the first stub was the maximum to be spent. All the checks were signed by Hille. When the pigs were bought, a copy of the bill of lading for their shipment and the checkbook would be mailed to the bank. The buyer would ride caboose and look out for the stock. In due time, the buyer would walk casually into the bank, and Hille would ask, "How much do I owe you?" There was not likely to be any argument; buyers did not ask for much.

A few weeks after the buyer had departed for South Dakota, where pigs were said to be plentiful and cheap, Father received a phone call from our Union Pacific depot agent. There were four carloads of pigs, consigned to him and Hille, and they were making a lot of noise. Later, Father found that there were between four hundred and five hundred pigs—lean, healthy, noisy and hungry pigs. The buyer had not disobeyed instructions. The pigs had departed from South Dakota in one freight car, triple-decked. The weather was hot and the pigs distressed, so the railroad agent at Kansas City had them unloaded, watered and reloaded into the four cars in which they had arrived in WaKeeney.

Father decided that he could feed half the pigs. He would sell the rest at auction after they had freshened up on barley pasture and the males had been castrated. There were plenty of farmers who would buy one to five pigs and feed them for meat for the family. I learned how noisy castrating pigs could be. Hired men did the surgery; chickens, ours or the hired men's, ate the excised gonads. Few infections or long-lasting disabilities resulted.

Feeding pigs with continuous access to self feeders and green pasturage and water was different from what I had expected. There was little "piggish" behavior; there was no crowding and biting. Typically, one or two pigs would be munching away at the feeders, with plenty of room to maneuver. They usually defecated in the pasture, so keeping the barn clean was not a problem. We ground a fifty-bushel load of wheat every few days, scooped it into the feeders and that was that.

The other lasting impression came later. As the pigs fattened, they took on a sleek look, but their sides seemed less convex than I had expected. When a hired man butchered one for our table, he skinned it, rather than scalding it and scraping off its hair as we had done in earlier years. Thus, a considerable coating of fat was removed from the meat. What was really distinctive about wheat-fattened pork was the flavor. It was neither bad nor good; it was *different*. A whole wheat porkchop sandwich did not need bread. The bread flavor was in the meat.

Father later said he felt he had got between fifty and sixty cents per bushel, instead of twenty-five, for the wheat. Probably he did not include overhead, such as wages for hired men or gasoline (about twenty cents per gallon) for grinding the wheat. What was important was that a few men had earned wages, we had had meat to eat, and all of us, including the pigs, had escaped boredom through a hot summer and fall and into the winter.

After Father's death, Mary found a carbon of a scathing letter Father had penned to his banker-partners in the Mason Garage venture. Father told of his sorrow that the bankers had stripped him of his most valued possession: the validity of his promise to pay. The specifics were not spelled out, but one could conclude that Father's scathing letter to his former partners arose from a request for financing to buy small lots of cattle and ship them to commission firms at Kansas City stockyards and their refusal to honor his note.

Later, the cashier at Hille's bank asked Mary to stop in. When she did, he presented her with a statement and canceled checks of a checking account in Father's name, an account no one in the family was aware of. The checks, mostly small, had been given for livestock to local farmers; the balance in the account was somewhere near a thousand dollars, and the cashier gave Mary a draft for that amount, closing the account. Thus it came to light that Hille had honored Father's name and given him an occupation for his last days.

The scathing letter and the checking account existed separately in my memory for more than forty years, drifting together only recently. They are in character. Father resented not being taken as a first-class credit risk, but though discouraged, he never quit trying.

Meanwhile, the cost of taxes, light bills, groceries and clothes were eating away at the Mason Garage account at the WaKeeney State Bank. Father soon came up with another project to recoup the garage's losses and the family fortune.

CHICKENS

It must have seemed unfair to Father that a dozen eggs would sell at the farm for as much as a bushel of wheat. This was true for a short time during the Depression, and this imbalance stimulated Father to go into egg production in a serious way. The Booth Hatchery, located at Clinton, Missouri, provided day-old chicks and detailed advice. Father's experience with chicken raising was limited to casual care of barnyard flocks, but he was eager to learn. Besides, activity—any purposeful activity—would keep psychological depression from piling on to the depressed state of the garage and implement business and of wheat farming.

With credit extended by Obe Northrup of the Verbeck Lumber Company, Father remodeled the farm's buggy shed into a brooder house and bought a kerosene burning brooder stove. He ordered five hundred white leghorn chicks from Booth's. They were scheduled to arrive at a time of year that would have them at laying age soon after January 1, when eggs traditionally were at their highest price. Booth's guaranteed fifty percent females, and added an extra one hundred chicks to the order so that two hundred and fifty pullets were assured. We followed directions as well as we could. We placed the brooder stove near the center of the room so that chicks crowding near it would not be smothered in room corners. We visited the brooder house two or three times each night to check on temperature and signs of unhappiness among the chicks. Nevertheless we had a few losses due to crowding. Apparently a few chicks would gather near a warm spot in the brooder canopy, others would crowd in, and with escape cut off on three sides, they would form a moist, suffocating mass. We learned by experience and raised more than two hundred pullets to egg-laying maturity. Cocks were sold off as soon as their sex was known—much later than more expert chick sexers would have sold them.

Father had provided nest boxes along the inside walls of the laying parlor, and, sometime before February 1, we were collecting more than two hundred eggs each day. Moreover, after a few weeks, the eggs were large. I remember that the gross weight of a thirty-dozen case of eggs had to be fifty-four pounds or more to be sold as *large eggs*. Our cases were soon averaging fifty-six or more pounds. Eggs were being bought for about twenty-seven cents a dozen at the cream and egg station in WaKeeney, and we felt that the chickens, like the pigs of the year before, were paying us for our labor. It was necessary, however, to feed the laying hens commercial feed. This ate up most, but not all, of the cost margin. Father attempted to substitute ground wheat, but even when it was mixed with ground bone and slaughterhouse scraps, this was not successful. The rate of egg laying did not drop, and the sizes and firmness of the eggs seemed to hold up. The trouble was that the hens developed an avid appetite for meat, which they found at each other's egg-laying apertures. Once a hen bled, she wouldn't last more than a few hours. We promptly culled all the injured hens, isolated the more aggressive hens and restored commercial feed. We got the trouble under control, but by the time we did, our case was, as a lawyer might say, moot.

When the price offered at the cream-and-egg station fell from twenty-seven to twenty-three cents, we held our cases of eggs in the tornado cave, a reinforced concrete structure mostly below ground level, about ten by twenty feet in floor area. It had been built by Hays Porter, who had made a good underground cellar.

We turned each case ninety degrees every day, to keep the egg yolks from settling to one side, and thus kept the product salable. The price recovered a cent or two, and we sold the eggs. But the trend was steeply down, and soon eggs sold for fifteen cents a dozen, then for a cent apiece and finally two for a penny. We could not buy commercial feed to keep the hens healthy and happy. I shot jackrabbits with my .22 Winchester Special rifle, and though I could keep a rabbit carcass available much of the time, it was no use. We took down the fences around the chicken runs, but the birds had been bred or educated, or both, to cluster together in and near the chicken house. Finally, we crated the hens and took them to town, taking whatever small price was offered. Some may have gone to farmers as gifts, where a dozen hens, living semiwild, might have survived and furnished eggs for the table, but there were few takers.

The depressed hens had turned into cannibals, but the depressed farm families, for the most part, turned to bootleg whisky and county welfare. Mary, who became County Welfare Officer, wrote up food orders for many of our old neighbors and garage customers. This was the social and economic climate into which Henry Erichs, like an angel of hope, came with an offer to buy the garage as a truck repair station. This relieved us, but Erichs, after a few years, was destined to die in the wreck of one of his trucks. Anyone who was on the road—truck drivers, bus drivers, even casual travelers—drove on our improved roads until they went to sleep and ended up, at best, with an injured vehicle but too often dead in a puddle of blood. Regulation and the return of profitable conditions helped reverse this deadly trend, but even today, heroic effort behind a steering wheel costs many lives.

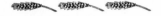

WEEKENDS

After we sold the garage property in December 1933, and I enrolled in Fort Hays State College in January 1934, our family—Father, Mother, Mary and I—continued to live in the fine two-story and basement brick veneer house in town that Father and Mother had bought from Sam Campbell at the time Father and the bankers formed the garage partnership. Often I stayed in Fort Hays for three to four weeks at a time. At other times I would board the Union Pacific's jitney train, (at first it was one coach and one express car, pulled by a steam locomotive, later it was a gasoline-powered single unit with a compartment for passengers and one for express packages and mail). It arrived in WaKeeney in mid-forenoon from points east, continued to Limon, Colorado, and returned, departing from WaKeeney for points east at two o'clock in the afternoon. It stopped, as farmers joked, "Whenever it saw a milk can on a station platform."

On many of these visits, I engaged in panic housecleaning, at first beating rugs hung on the clothesline so that their patterns emerged from the plain brown of the dust with which they were saturated. Later—rugs rolled up in the ends of the rooms—I swept and shoveled dust that had beaten in under doors, around window frames, seemingly even through the walls themselves. Trips to the farm, to which my parents still held title, were depressing. In the twenty years before Father had entered the garage and implement business, he, like many other Kansas farmers, had anticipated the New Deal's shelter belt program, planting rows of Chinese elm, box elder and red cedar trees to grow into windbreaks sheltering the farmsteads, in our case on the west, north and south. A well had been drilled northwest of our farmhouse, and a windmill put over it to pump water almost continuously from spring to autumn down hoe-scratched runs between rows of trees and to rows of potatoes, melons and other vegetables for the table. Periodically, the spaces between tree rows outside the garden patch were cultivated to keep what Father called a dust mulch loose over the subsoil. The idea was that this would absorb any rain that fell and, between rains, slow the evaporation of moisture from the soil below the depth of cultivation. In those earlier days, dust did not blow to any great extent. Whatever the merits of the method, for twenty years the trees prospered. The elms and box elders and a few locust trees made an almost continuous shade by 1925, and the interlocked red cedars had sheltered the house until drifts of winter snow, forming between trees and house, dampened the soil enough to make my parents dream of a real lawn west of the house.

In 1934 and the years immediately following, the drifts of snow had been replaced by drifts of dark brown dust, towering as high in places as the fifteen-foot ridge of the house roof. There was a heart-wrenching difference, however. Snow had always lofted over the cedars and settled beside them, the tops of the drifts several feet to the south or east of the trees. The drifts of dust had centered themselves on a line through the rows of tree trunks, completely burying and choking the cedar branches. One or two spindles of scraggly stem might extend a foot and a half above the dirt. There was no way to remove the dirt without also taking the trees. Why not sculpt the slopes of the drifts and plant grass or other ground cover? I have read that water erosion is more destructive than wind, because blown dirt eventually settles, while washed dirt often ends up skittering over the sands in creek bottoms, accumulating in river deltas or the Gulf of Mexico. The sad truth is that drifts of dust are not soil in any real sense. I remember being bored in an agriculture course by a seventh-grade teacher who read us a paper, perhaps her master's thesis, detailing all the kinds of water found in a soil sample, distinguishable by the way each kind adhered to walls of dirt or moved up and down through the soil profile. A dozen years later, I could see what her essay had been all about. Soil is structured. Dust drifts are not. Dirt scooped up to make an artificial lake in a park in Walla Walla, Washington, could be worked into berms and, years later, covered with grass. Our children loved to climb them and slide down them. But the bulldozers that heaped up these berms did not completely destroy the capillary spaces and the debris of roots. In a few years, the soil had forgiven its rude disruption and was cheerfully producing grass and flowers. Not so the Kansas dust drifts. Father sadly hired a man with a road grader and a bulldozer and removed dust, trees and all, from our cedar windbreak. I do not know where he put this refuse, probably in a pair of dugout remnants that had long been part of the pig pasture. They had been, my parents said, temporary shelters for people named McCormick, who had proved up the land as a homestead. (My sisters and I once found a man's gold ring among the bits and shards there.) Whatever the fate of the dust drifts, I was not there to see it. Father and hired men carted away the lath and plaster from walls and ceilings of the farm house. Dust drifting through shingles and under eaves had brought ceilings down. The rooms were repaired and, I remember, oak floors were laid on top of the pine boards we had known as children. The oak came from salvage of the showroom floor after the garage had burned down in 1929.

No one who knew them could say my parents were prone to quit. When my wife, Isabel, visited them in the summer of 1941, while I was at Purdue finishing my Ph.D. thesis, Father and Mother were living at the farm. They had sold the brick house, and were waiting out the remodeling of a smaller one in WaKeeney, where they lived until Father's death in 1944.

In the mid-1930s, there were always chores for me when I spent a college weekend at home. One time it was to unlock the garage safe, which had been trucked to the brick house and left on the front porch. It had a four-digit combination, and both Mary and I knew the numbers and their proper order. One had to twirl the dial clockwise, stopping at the first number. Then one must "pass" the first number twice, turning anticlockwise to the second number. Then one had to turn the dial clockwise, pass the second number once, stop at the third number, then turn the dial anticlockwise directly to the fourth number. Mary "knew" all this, but was too jittery to count the passes correctly, so the safe remained locked. Fresh from school, I had no trouble, unlocking the safe on the first try. No one put anyone down or gloated.

On another occasion, Father took me several miles down a country road to inspect a McCormick-Deering tractor he had traded for. The farmers who had owned it had bought new cylinder sleeves, pistons and so forth; had the valves properly reseated; and then, unable to start the machine, had given up and taken their loss, trading for a John Deere Model D, perhaps new, perhaps used and serviceable. Here I had several advantages over the farmers who had done the work. First, I had had more experience; second, I had not worked myself into a frenzy of frustration; and third, only Father and I were there. There was no gallery of onlookers. A check showed that valves had approximately the right clearance. The #4 exhaust closed just after top center, but the #1 spark plug failed to fire when the magneto impulse tripped. Instead, a thin wisp of smoke came from the stack. The wrong cylinder was receiving the spark. It was obvious that the magneto had been taken apart and put together with its distributor gear improperly meshed. It was not hard to disassemble the magneto far enough to remove the distributor gear and place its brass segment so that it would lag the proper cap segment a bit, rather than to face it full on when the impulse starter tripped. This corrected, we cranked the tractor and I drove it home to the brick house. Father was more than pleased. Throughout this period, most tractor manufacturers had, starting with horsepower ratings of 15-30 (drawbar-belt) gradually inched them up by increasing engine speeds, overboring blocks and raising compression, until they could advertise larger horsepower numbers than their competition. McCormick-Deering had the "best" numbers going, so its tractors sold well, new or used. John Deere, hampered by a less flexible two-cylinder design, could not compete in the numbers game, and its products appealed only to the conservative values of simplicity, long life and ease of repair. It seems that *bigger is better* touched something precious and primitive, dear to the Kansas farmer's soul.

EPILOGUE

At the Mason Garage, we hired a number of mechanics, none of whom had Paul Clem's way with automobiles. This did not matter greatly because Highway 40 was soon moved to the south part of town and tourists seldom brought us their overheated Packards and other monsters of the time. Business was passable but not brisk for the years 1930 and 1931. Then, in a matter of months, life became grim. Wheat prices went as low as twenty-five cents a bushel, hogs and cattle were equally cheap, and business life in WaKeeney came to a near standstill. In winter we no longer heated the shop area but this did not matter; there was no work. Father used a small gasoline heater in his cubbyhole office and, since the heater had no vent, it used up the air in the room and made a high sharp stink. Father tried all the schemes he knew to make money and some that he knew less than well.

Summer of 1933 passed into winter. The sharp stink of Father's office heater came out the door to meet us. We wanted to weld up a wrecking crane to go into a short-wheelbase truck we had but could not afford oxygen or acetylene. Into this gloom walked Henry Erichs, a local young man who had been running a truck line from Kansas City to Denver for several years.

193

The congealed bureaucracy of the railroads had made an opportunity for men like Erichs, who would cheerfully pick up a package of any size and drop it off at any town along the six hundred and forty miles of highway; his truckers employed a minimum of paperwork, and delay was not a problem. WaKeeney was a good point for a service shop for the Erichs line. It was within a few miles of half-way between the two terminal cities, and it was home. Erichs wanted to buy a place where he could store and service equipment. Father and Erichs measured our big overhead door and Henry decided it was tall enough and wide enough. They completed a deal about Christmastime. This time the banker partners were ready to let go.

Father's entanglement with the John Deere Plow Company was harder to unravel. The company's policy had been to extend free credit for the purchase of both tractors and other machinery for a current crop year. For unsold items carried over, they required notes bearing eight percent interest until invoices were cleared and notes paid in full. The policy was well calculated to prevent dealers from becoming rich, even in moderately good times. There were long negotiations between Father, his lawyer and men from the plow company. The men from Kansas City blinked first. By arranging inventories, statements and payments in chronological order, Father's lawyer, William Wagner, showed that payments that should have been credited against a note account drawing eight percent interest were applied, instead, to overpay a parts account which was not in arrears. In a hallway near the lawyer's office, the Kansas City representative of Deere asked Father how much he thought he really owed and was willing to pay to wipe the books cleans. Reaching into his still facile mind for a number, Father gave an amount of a little more than eight hundred dollars—the wholesale price of one Model D tractor. The plow company man extended his hand, Father shook it, the lawyer drew up the proper papers, and Father went across the street to alert his banker partners, who were more than happy to make sure that his check cleared. Father's farmland was soon sold to satisfy mortgages.

I know little about what followed; I was away at school. With the business sold and no occupation for me, Father saw that it was time for me to get more education. I could enroll in Kansas State College at Fort Hays for less than fifty dollars for fees and books; room rent would be less than thirty dollars a month, and food might cost a dollar and a half a day. Low college tuition, Roosevelt's youth programs and money from Mary's bookkeeping job were to ensure my future. Father suggested that I study law. For the first time in my memory I stood up to him. I did not know what I wanted to become only that I wanted to learn. I told him that I could not promise to study law. He did not protest. I enrolled for the spring semester at Fort Hays in 1934.

The battle with Deere was over, but the physical dust did not settle. Storms that blotted out the sun and turned day into night swept the plains, rolling across the land as visible, roiling walls of dust, suffocating cattle and burying mortgaged tractors, plows and wheat harvesting combines, sometimes completely. The air cleaner on a pipe above a combine was about twelve feet off the ground; sometimes it was the only part visible above a dust drift. Sometimes it, too, disappeared. Young men hired themselves out to government livestock buying programs and bought up the thin cattle left on farms whose owners had had to apply for public assistance. Even before the garage was sold, old horse buyers who counted Father as friend had been in the habit of sitting around the small gasoline-powered stove in the garage office, telling stories and soothing themselves with rotgut whiskey in the evenings. During the days they roamed the country offering cheap bed blankets and a few dollars for any horse strong enough to stand the ride in a truck to a skinning and rendering plant. Father still had title, held jointly with Mother, to the two-story brick veneer house on a good street in WaKeeney, a much-abused Dodge pickup truck, and friends, desperately poor but able to ply him with bootleg whiskey. The product was called *Bolshevik*, and the source of supply was presumed to be in Nebraska, about a hundred miles north of WaKeeney. Father roamed the country near WaKeeney, buying small lots of live-stock and bunching it for local livestock sales and occasional carlots to be shipped to Kansas City. Prices of real estate began rising, apparently in response to the threat of World War II, which would soon overtake Roosevelt's New Deal. Father arranged to sell the brick house and buy a much smaller one for himself and Mother in a respectable but "cheaper" part of town. Mother looked the place over, saw that lights came on with pull-chains instead of wall switches and refused to sign the necessary papers until Father promised to correct this and a few other deficiencies in the place. It was to this house that Mary called me from a job at the University of Texas in April of 1944. Father had died, probably in the cab of his truck, and probably drunk, on April 25. He was sixty-nine. Before prohibition, before the garage venture, before dust storms and the Depression, he might have taken a drink of whiskey when offered in a social setting, but he did not have a drinking problem. After the funeral, the family doctor, who had brought me into the world, apologized for not being able to keep Father alive; an acquaintance, perhaps ten years younger than Father, gave the condolence that the County had lost a good Republican. Perhaps. I think Father voted for Roosevelt at least once.

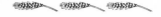